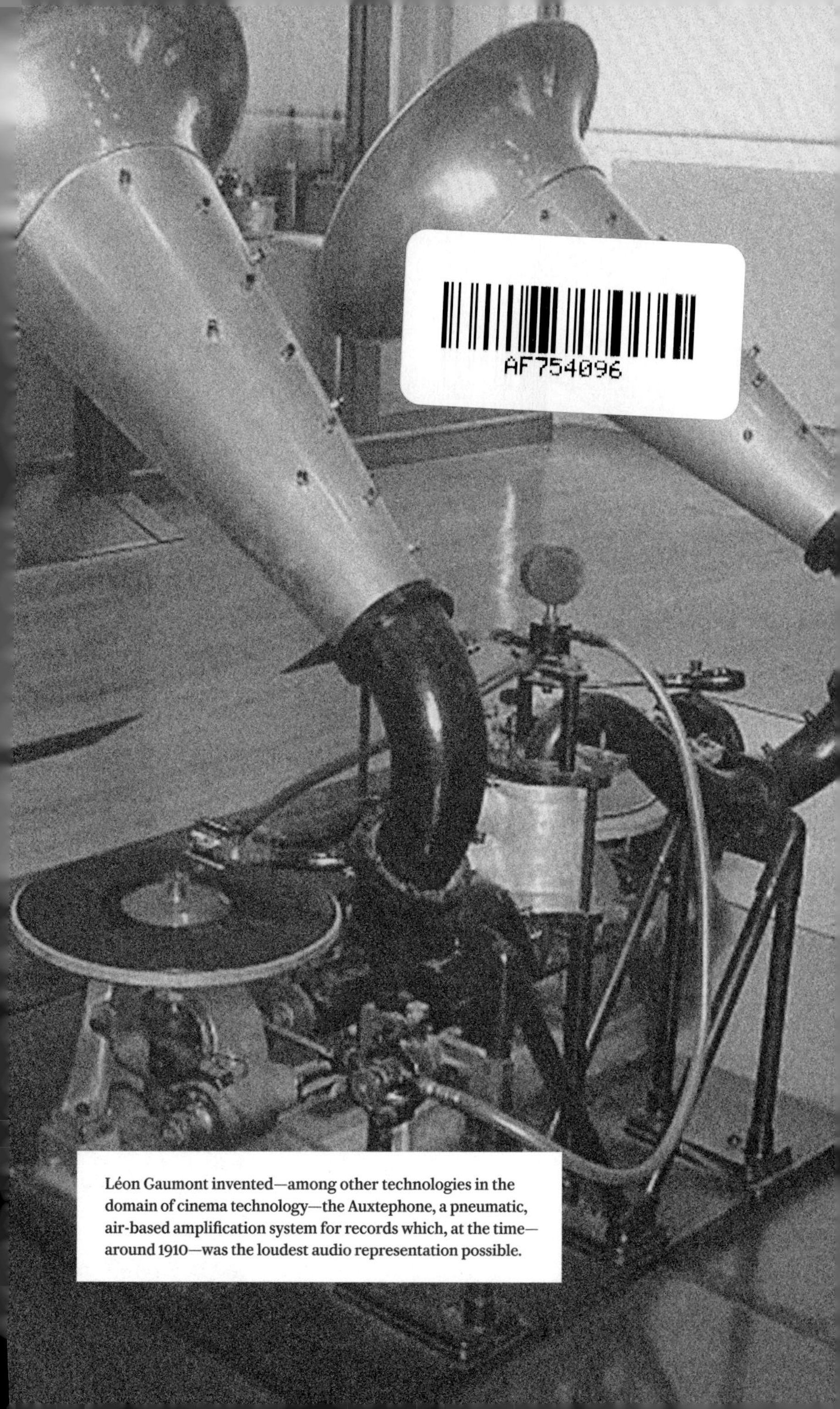

Léon Gaumont invented—among other technologies in the domain of cinema technology—the Auxtephone, a pneumatic, air-based amplification system for records which, at the time—around 1910—was the loudest audio representation possible.

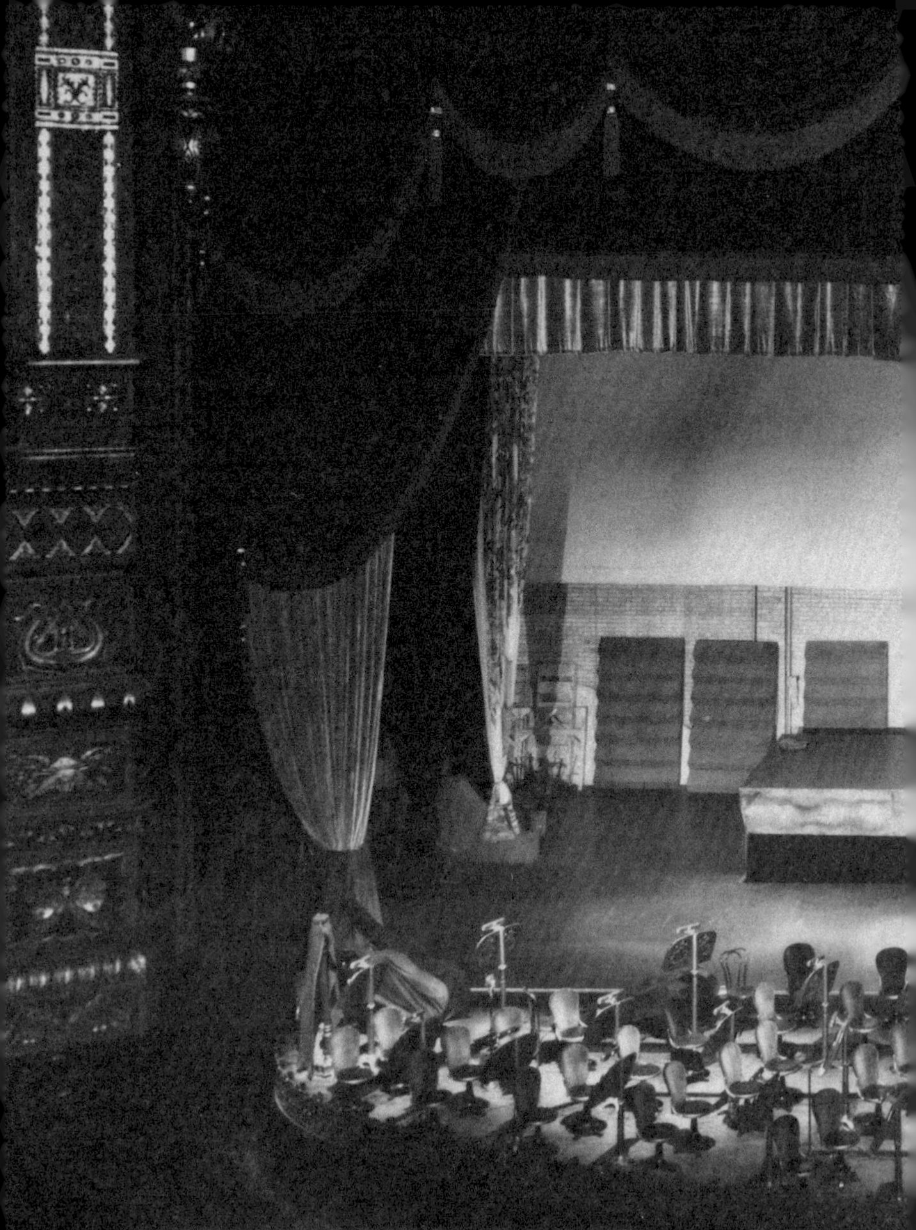

The exoticist Fox Theater in Detroit had an enormous mono audio system housed in the stage pit. The theater opened with this system in 1928, at the very advent of audio in cinema. At the time, this was one of the largest audio situations in the world, bringing together five thousand people at once.

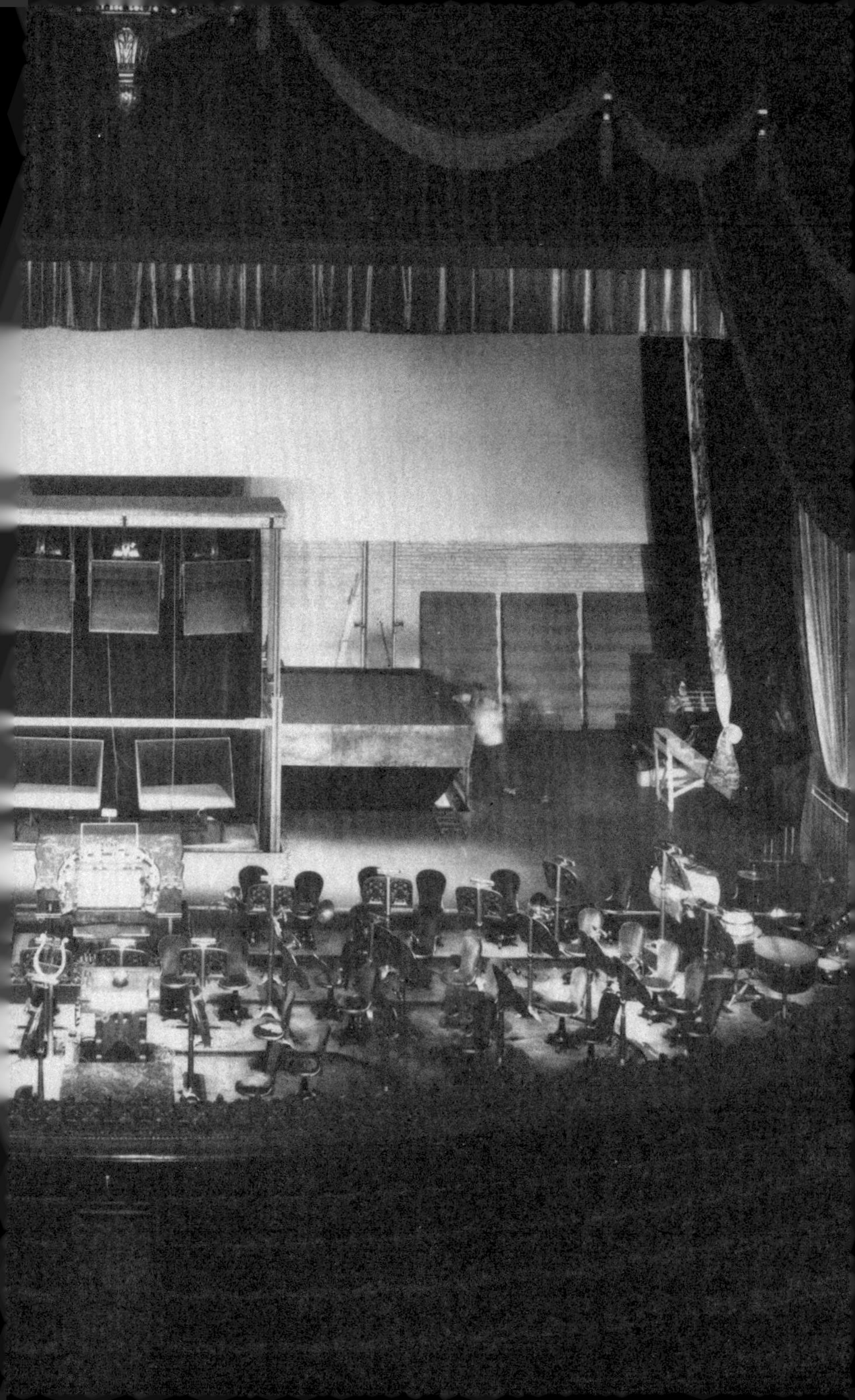

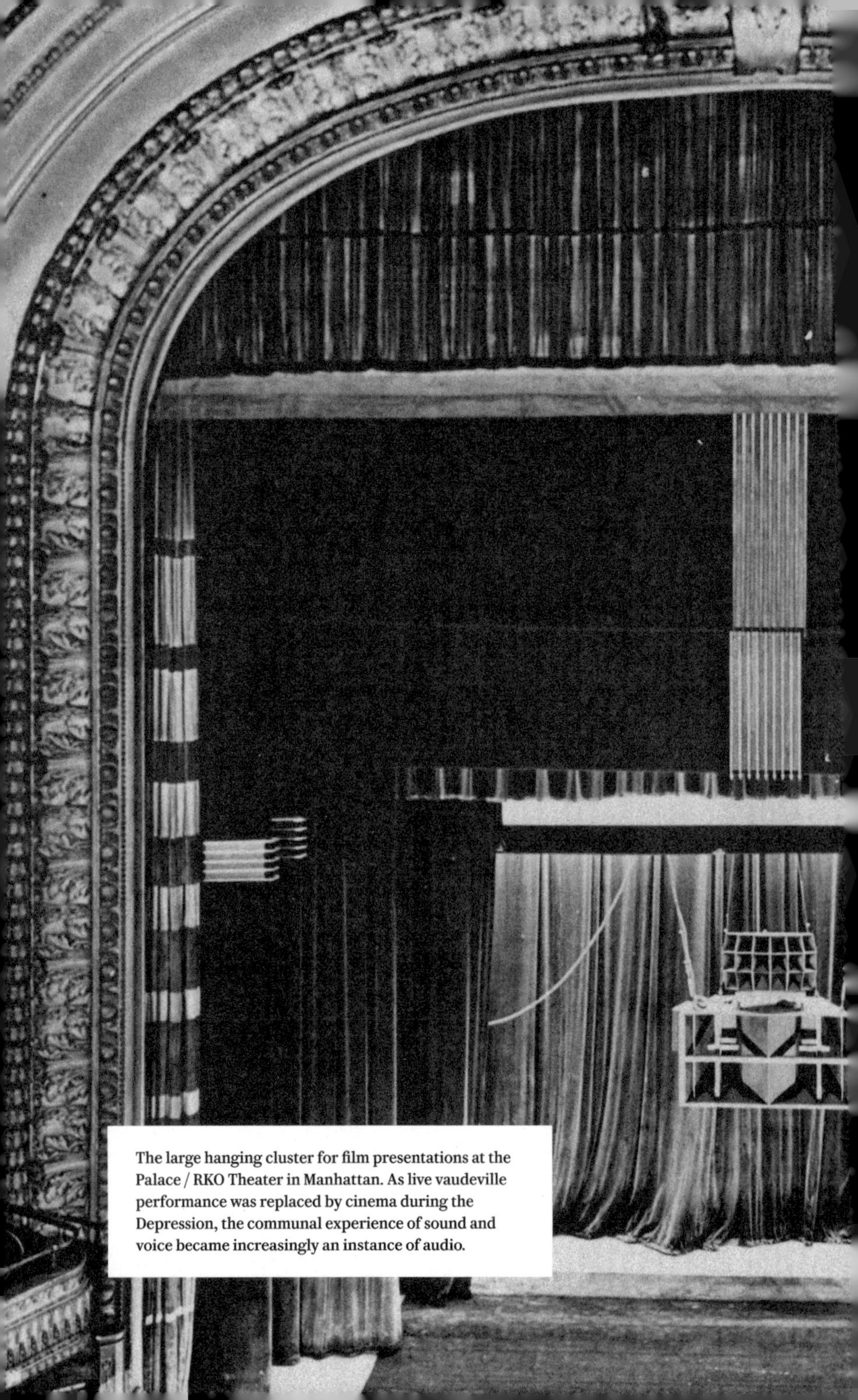

The large hanging cluster for film presentations at the Palace / RKO Theater in Manhattan. As live vaudeville performance was replaced by cinema during the Depression, the communal experience of sound and voice became increasingly an instance of audio.

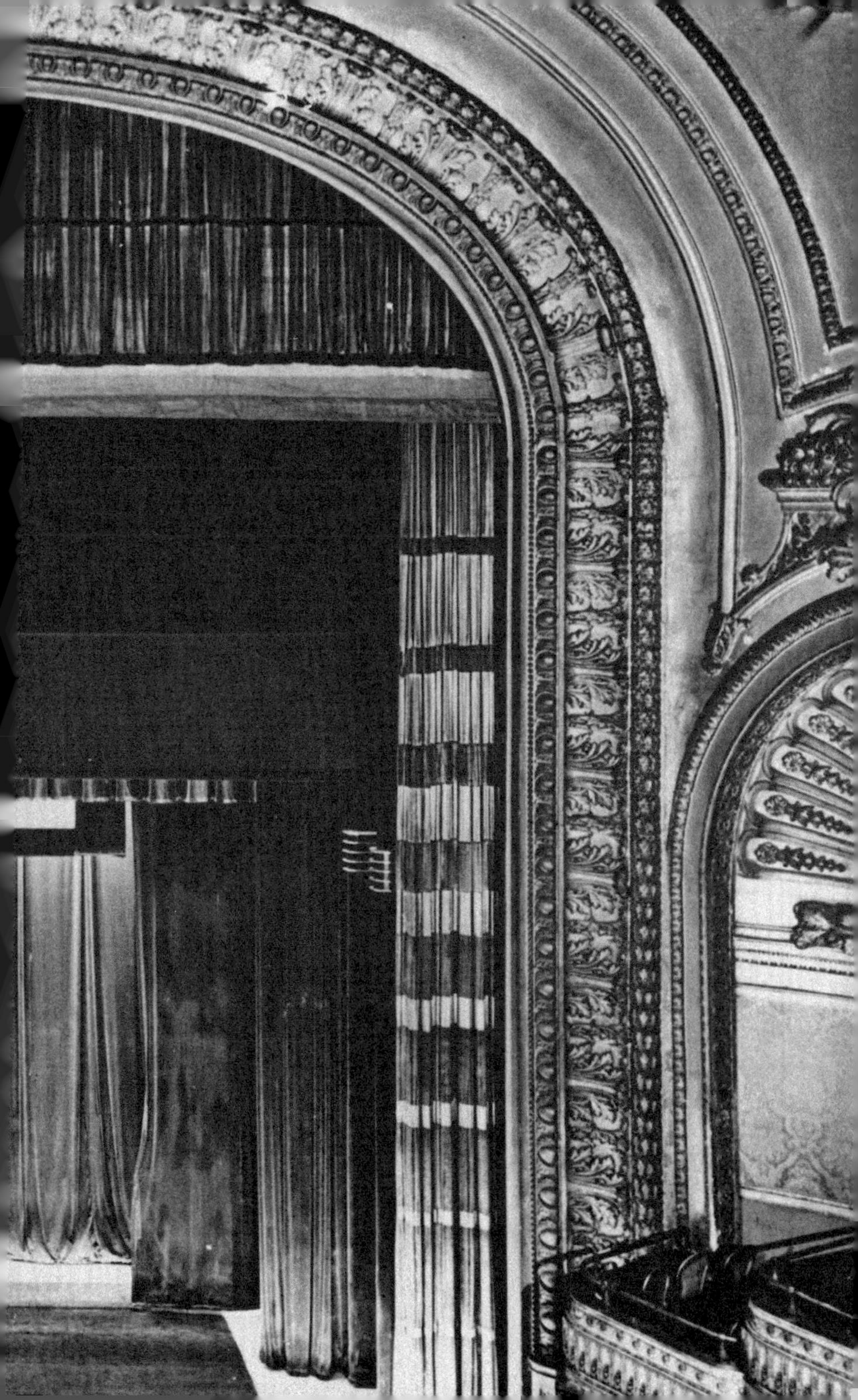

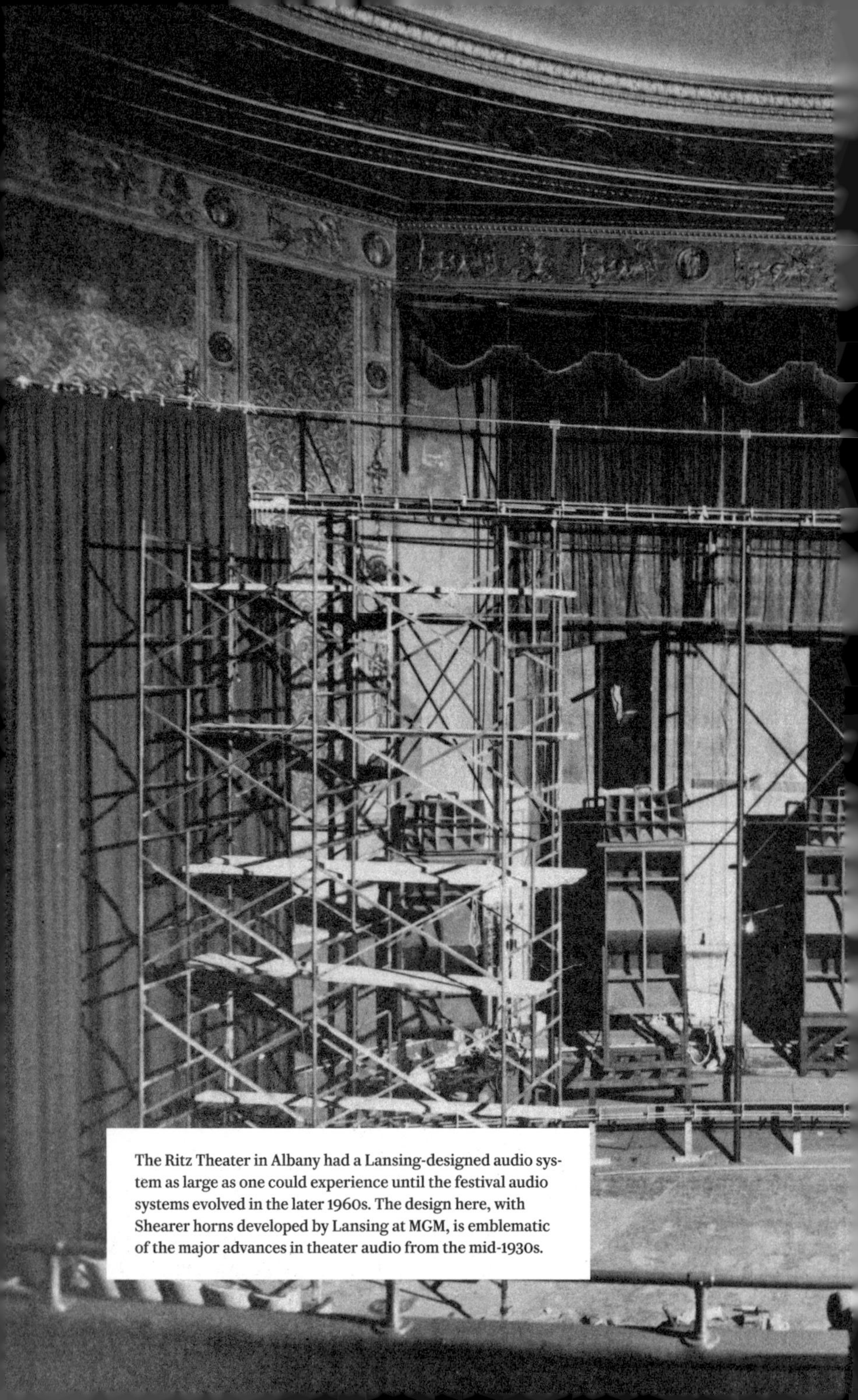

The Ritz Theater in Albany had a Lansing-designed audio system as large as one could experience until the festival audio systems evolved in the later 1960s. The design here, with Shearer horns developed by Lansing at MGM, is emblematic of the major advances in theater audio from the mid-1930s.

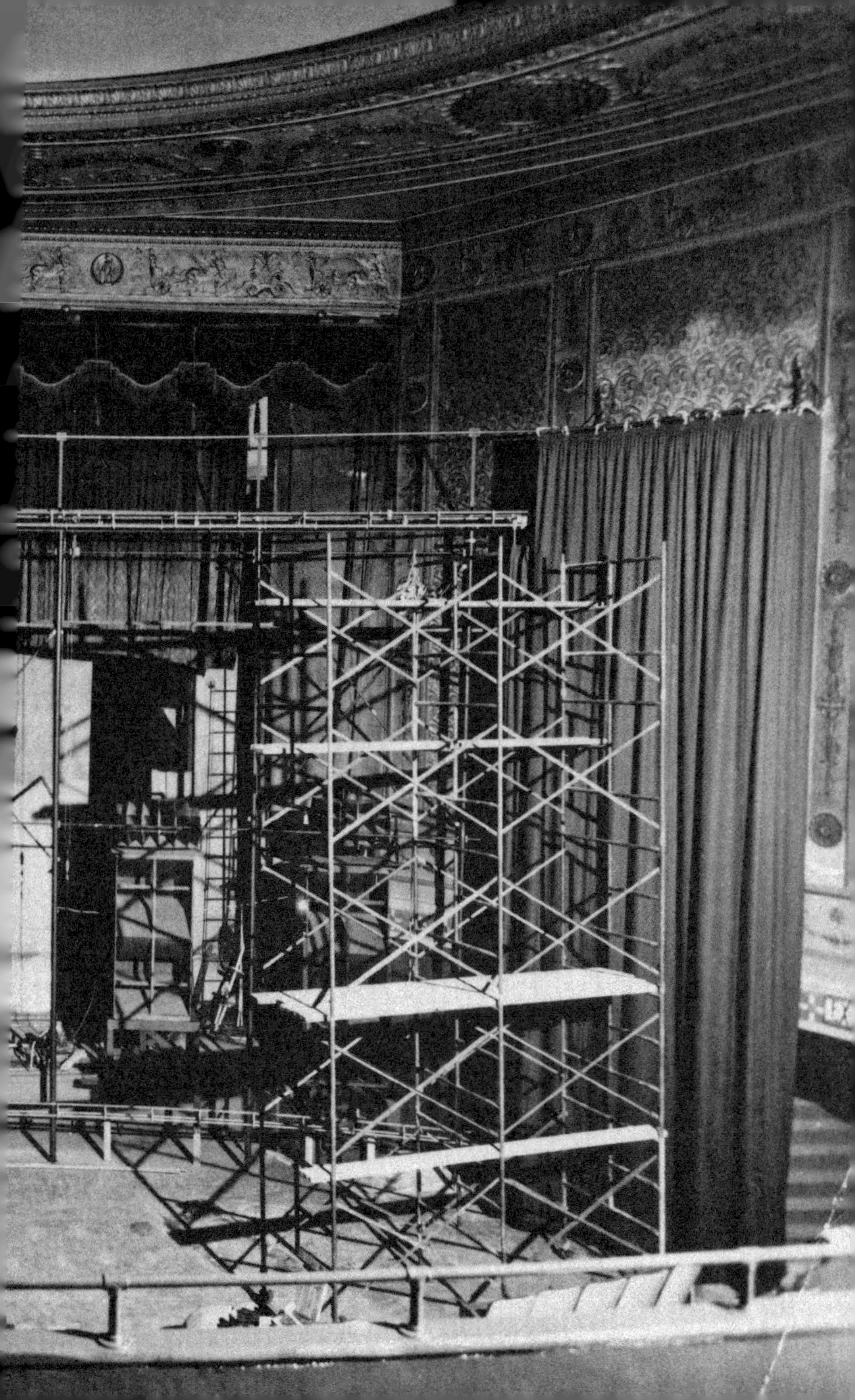

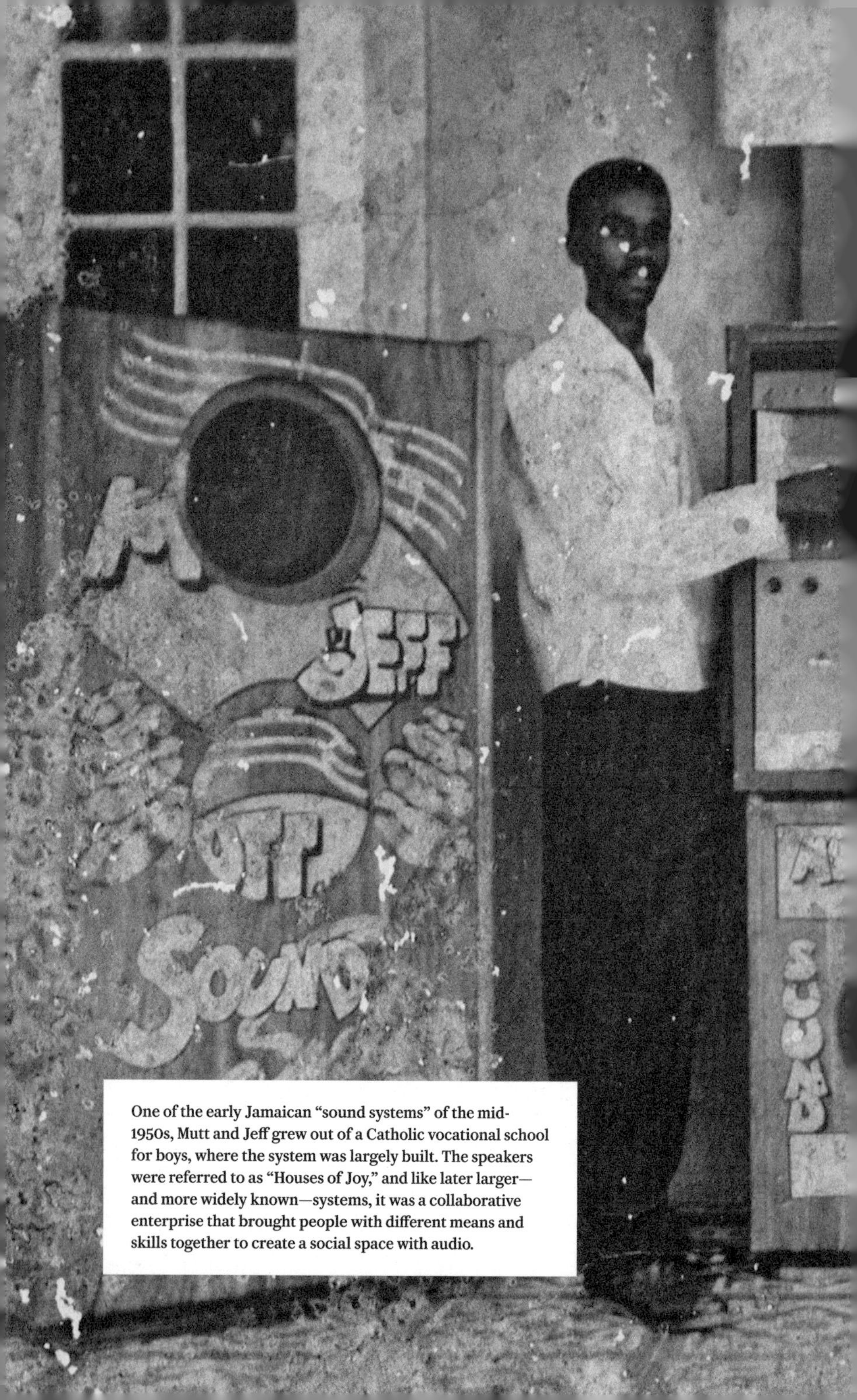

One of the early Jamaican "sound systems" of the mid-1950s, Mutt and Jeff grew out of a Catholic vocational school for boys, where the system was largely built. The speakers were referred to as "Houses of Joy," and like later larger—and more widely known—systems, it was a collaborative enterprise that brought people with different means and skills together to create a social space with audio.

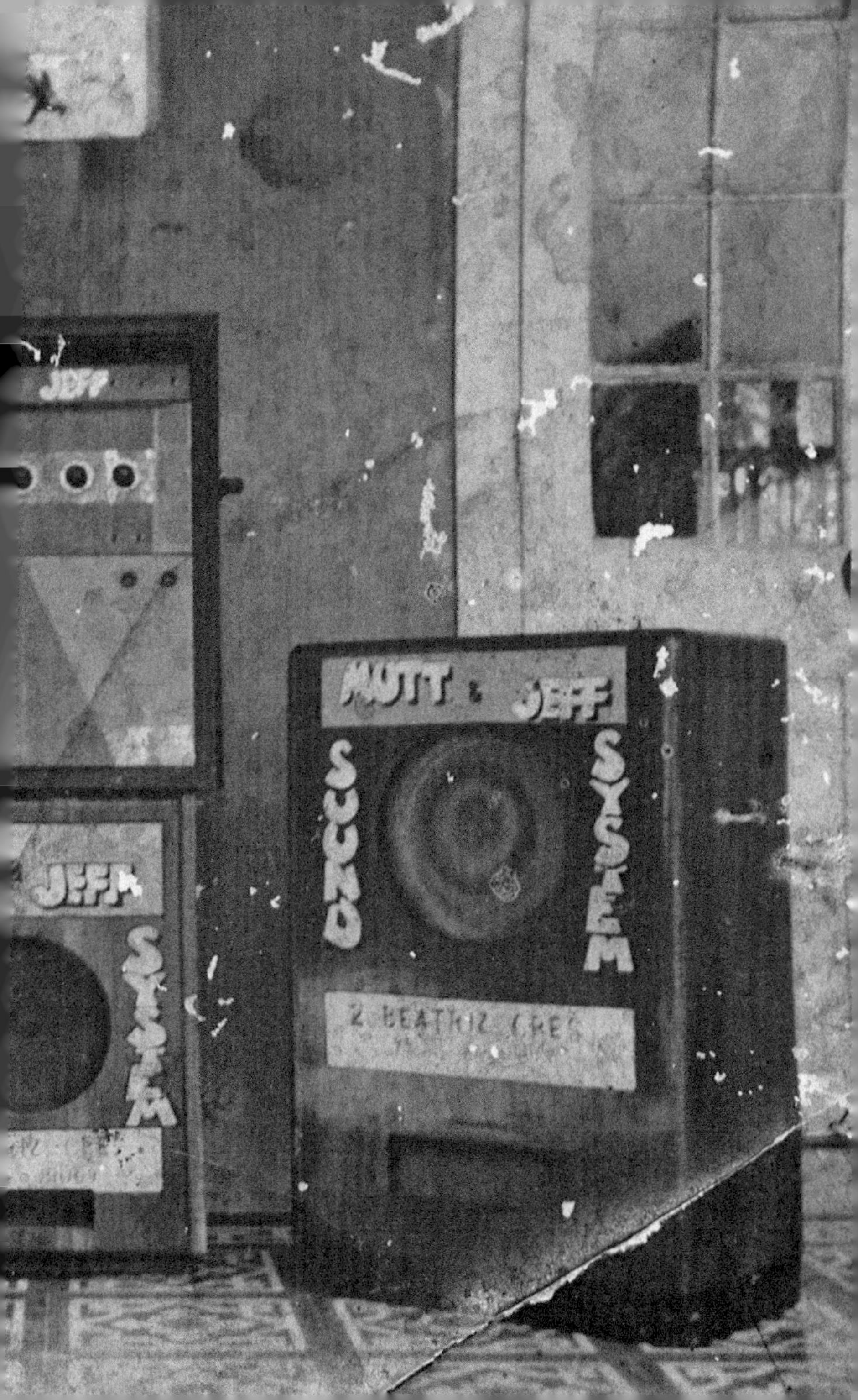
MUTT
JEFF
SOUND
SYSTEM
JEFF
SYSTEM

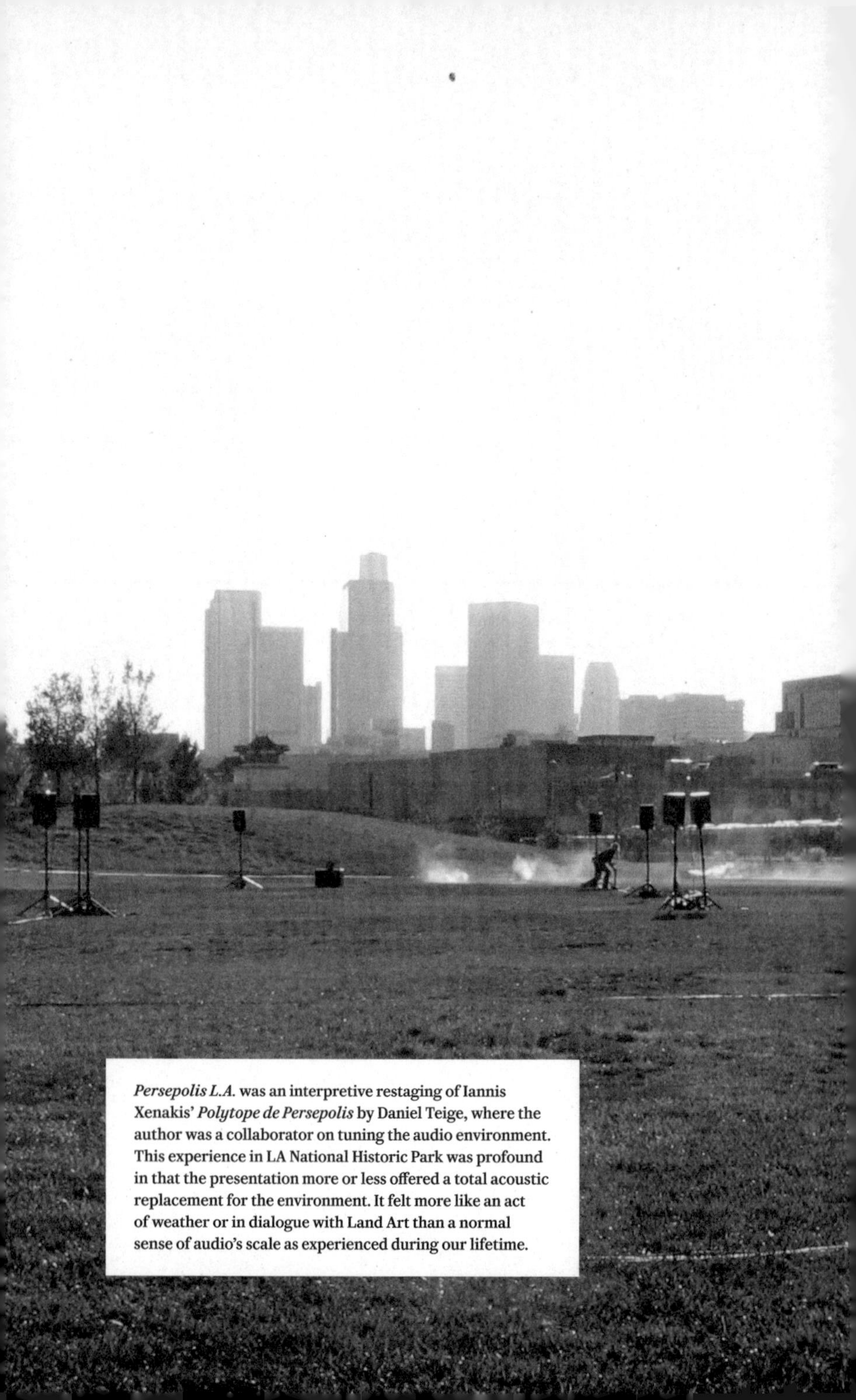

Persepolis L.A. was an interpretive restaging of Iannis Xenakis' *Polytope de Persepolis* by Daniel Teige, where the author was a collaborator on tuning the audio environment. This experience in LA National Historic Park was profound in that the presentation more or less offered a total acoustic replacement for the environment. It felt more like an act of weather or in dialogue with Land Art than a normal sense of audio's scale as experienced during our lifetime.

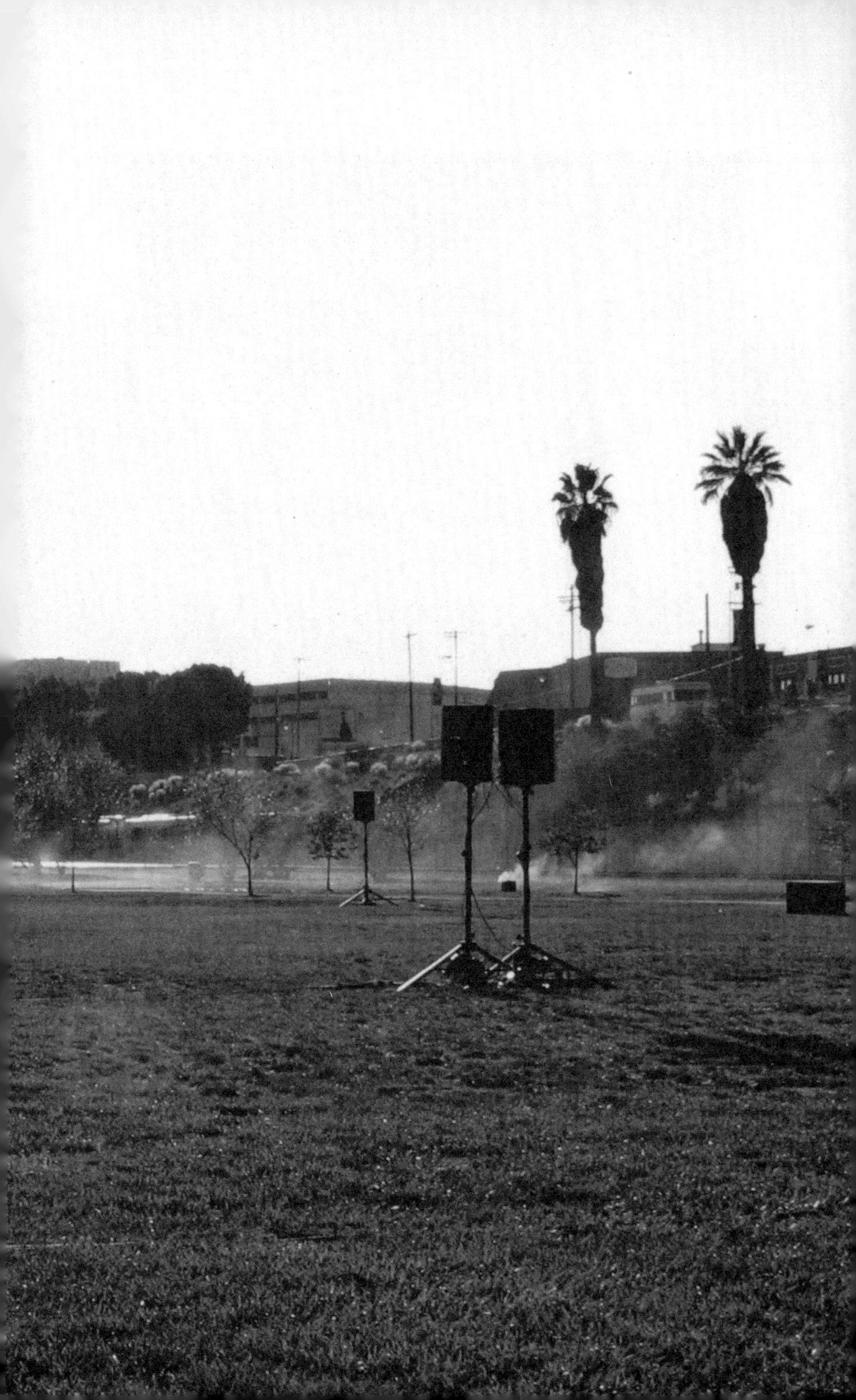

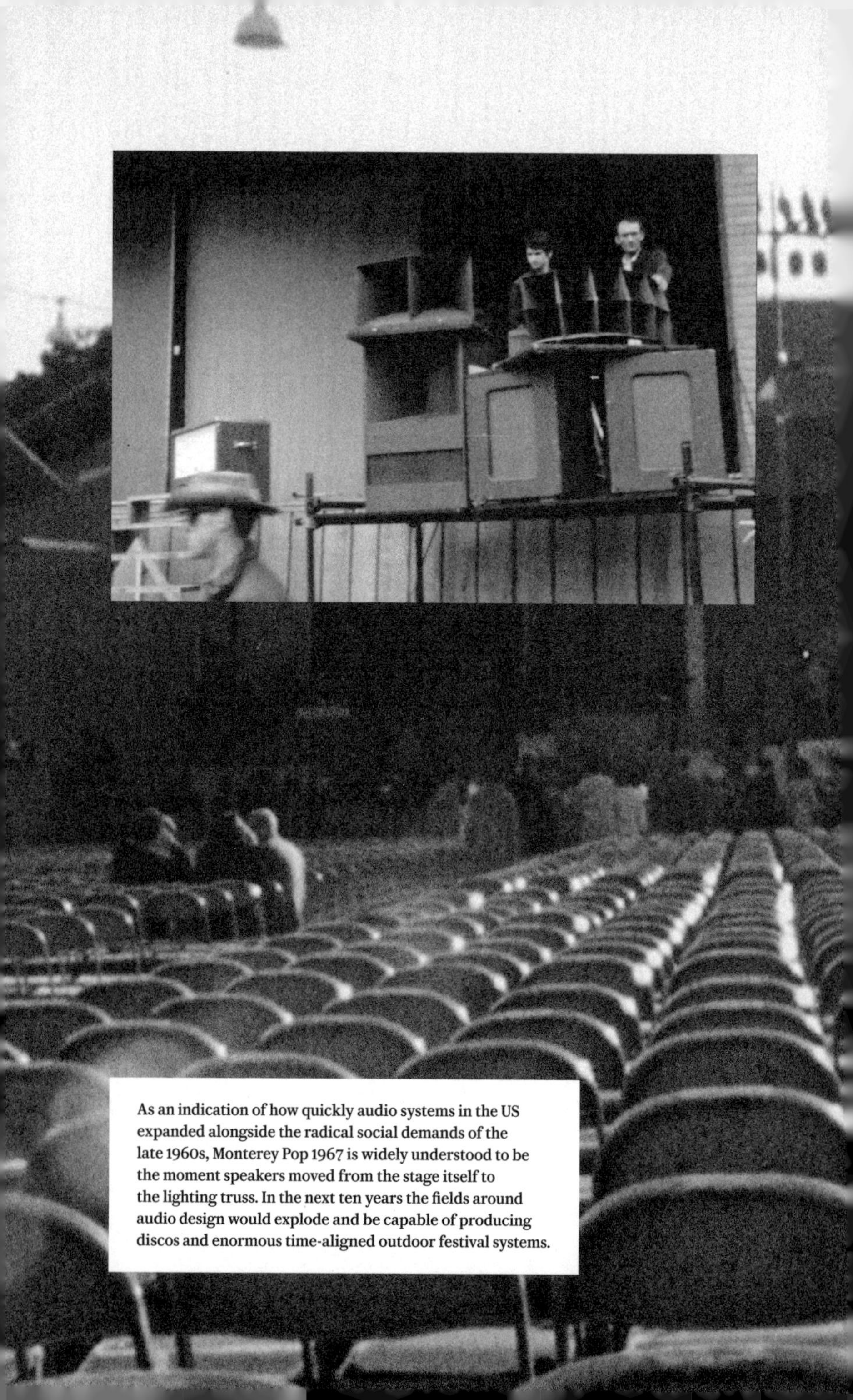

As an indication of how quickly audio systems in the US expanded alongside the radical social demands of the late 1960s, Monterey Pop 1967 is widely understood to be the moment speakers moved from the stage itself to the lighting truss. In the next ten years the fields around audio design would explode and be capable of producing discos and enormous time-aligned outdoor festival systems.

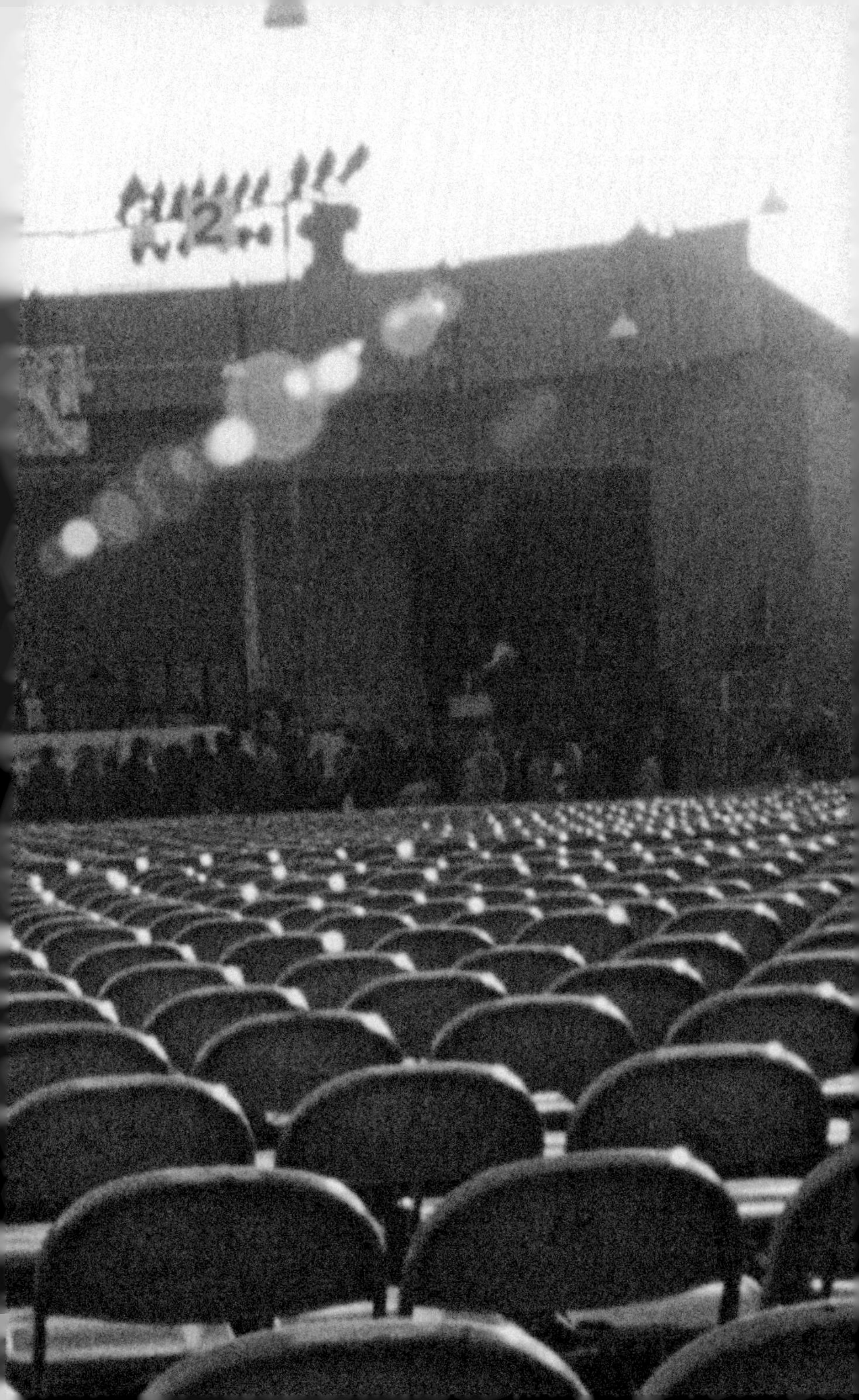

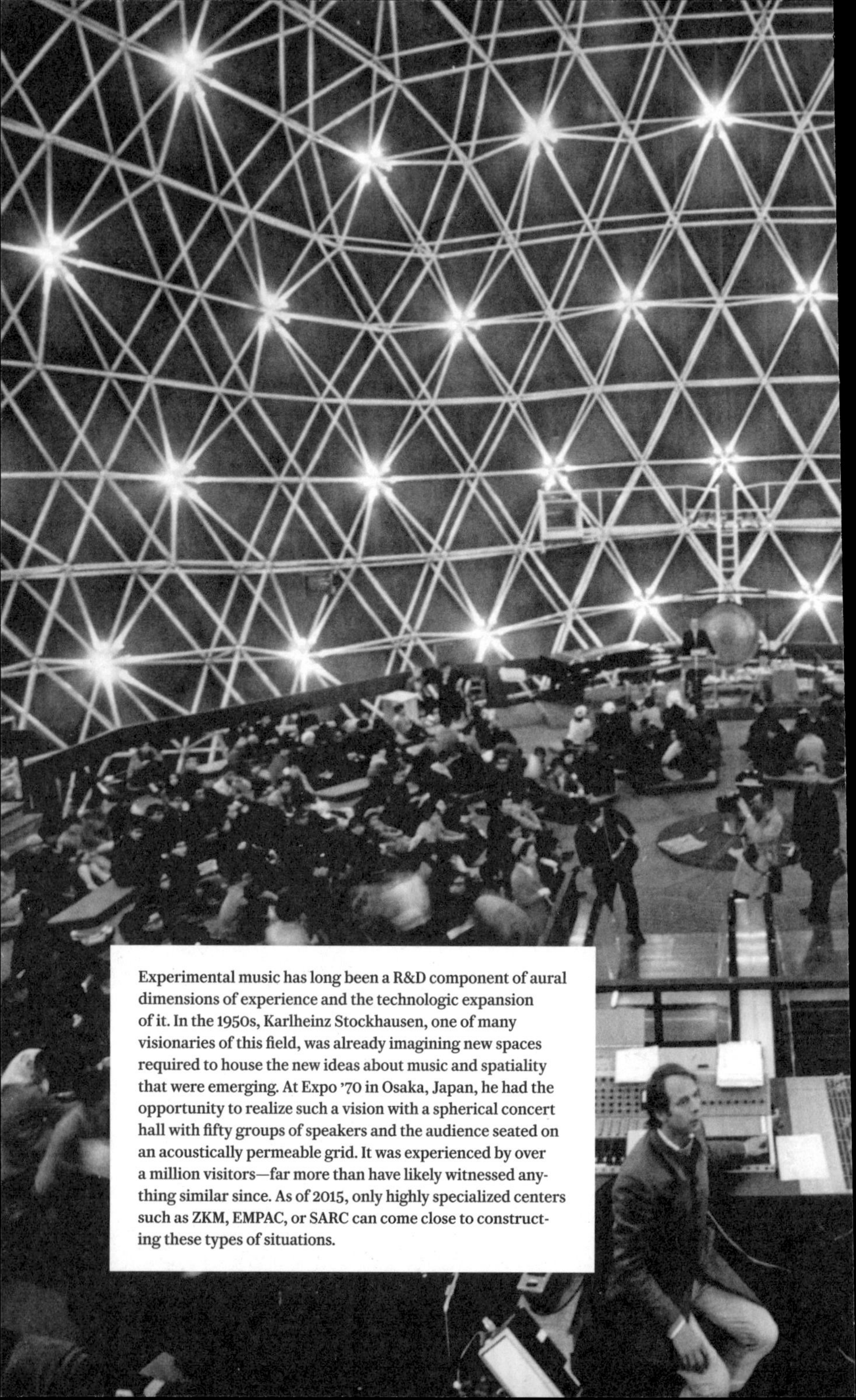

Experimental music has long been a R&D component of aural dimensions of experience and the technologic expansion of it. In the 1950s, Karlheinz Stockhausen, one of many visionaries of this field, was already imagining new spaces required to house the new ideas about music and spatiality that were emerging. At Expo '70 in Osaka, Japan, he had the opportunity to realize such a vision with a spherical concert hall with fifty groups of speakers and the audience seated on an acoustically permeable grid. It was experienced by over a million visitors—far more than have likely witnessed anything similar since. As of 2015, only highly specialized centers such as ZKM, EMPAC, or SARC can come close to constructing these types of situations.

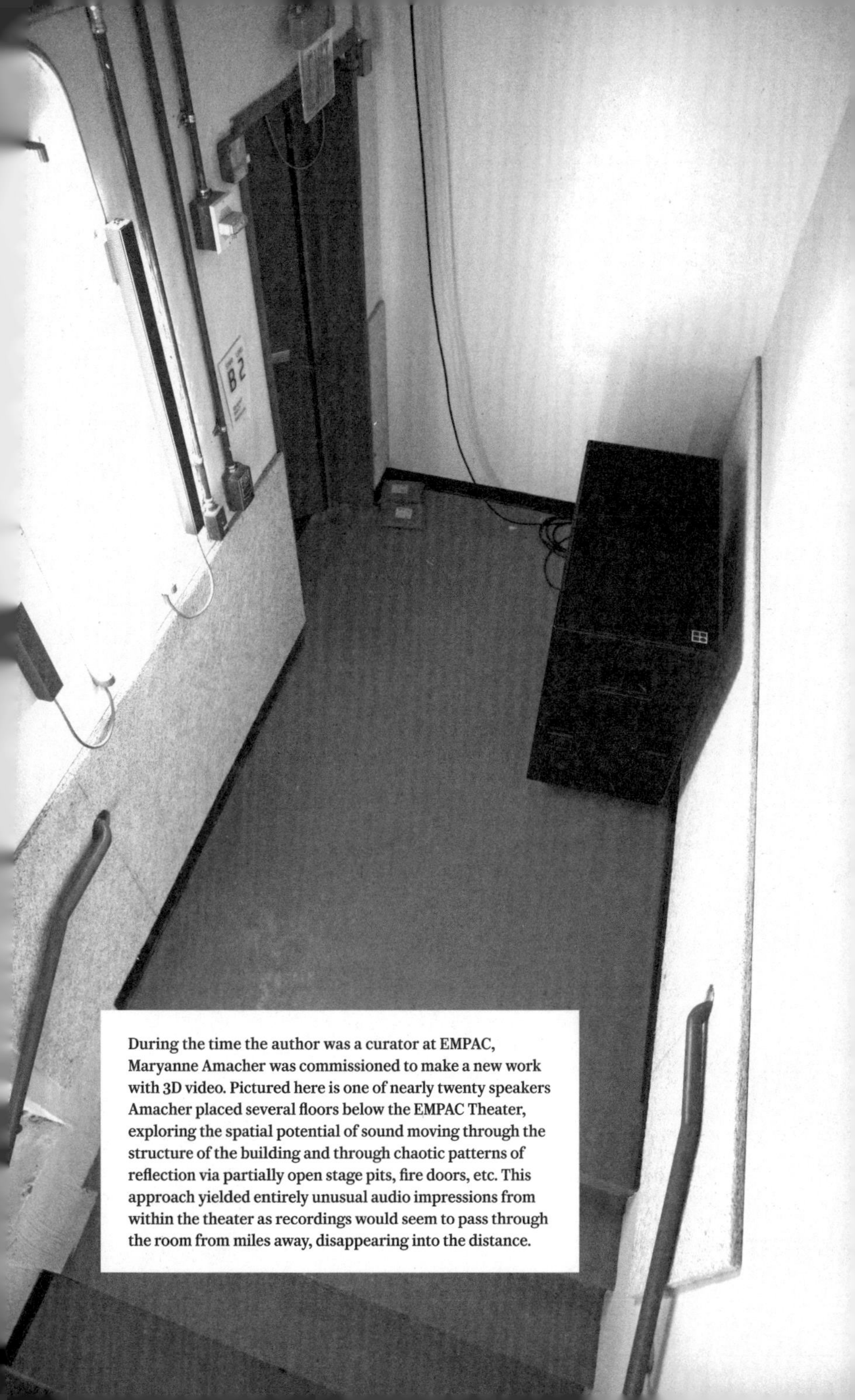

During the time the author was a curator at EMPAC, Maryanne Amacher was commissioned to make a new work with 3D video. Pictured here is one of nearly twenty speakers Amacher placed several floors below the EMPAC Theater, exploring the spatial potential of sound moving through the structure of the building and through chaotic patterns of reflection via partially open stage pits, fire doors, etc. This approach yielded entirely unusual audio impressions from within the theater as recordings would seem to pass through the room from miles away, disappearing into the distance.

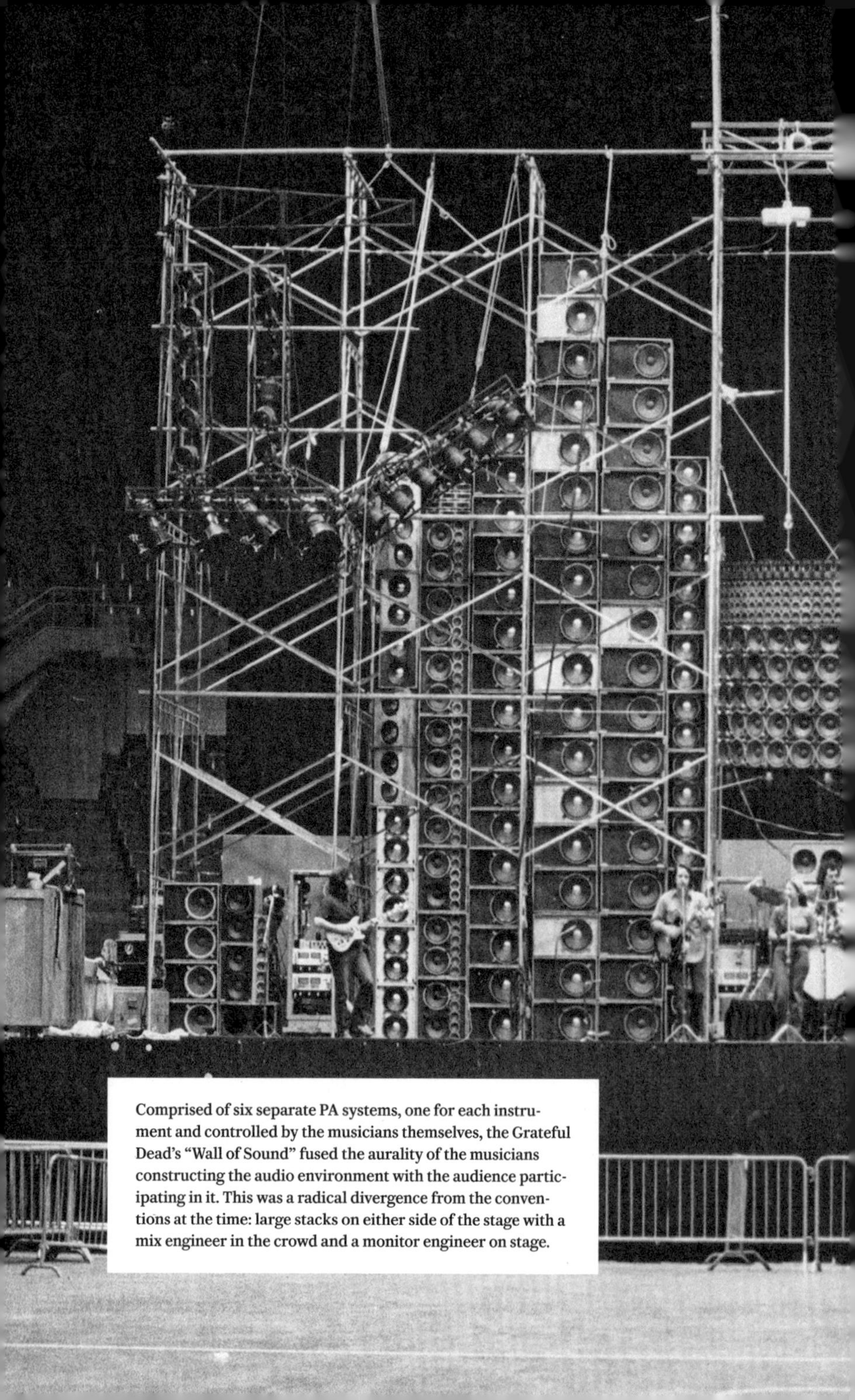

Comprised of six separate PA systems, one for each instrument and controlled by the musicians themselves, the Grateful Dead's "Wall of Sound" fused the aurality of the musicians constructing the audio environment with the audience participating in it. This was a radical divergence from the conventions at the time: large stacks on either side of the stage with a mix engineer in the crowd and a monitor engineer on stage.

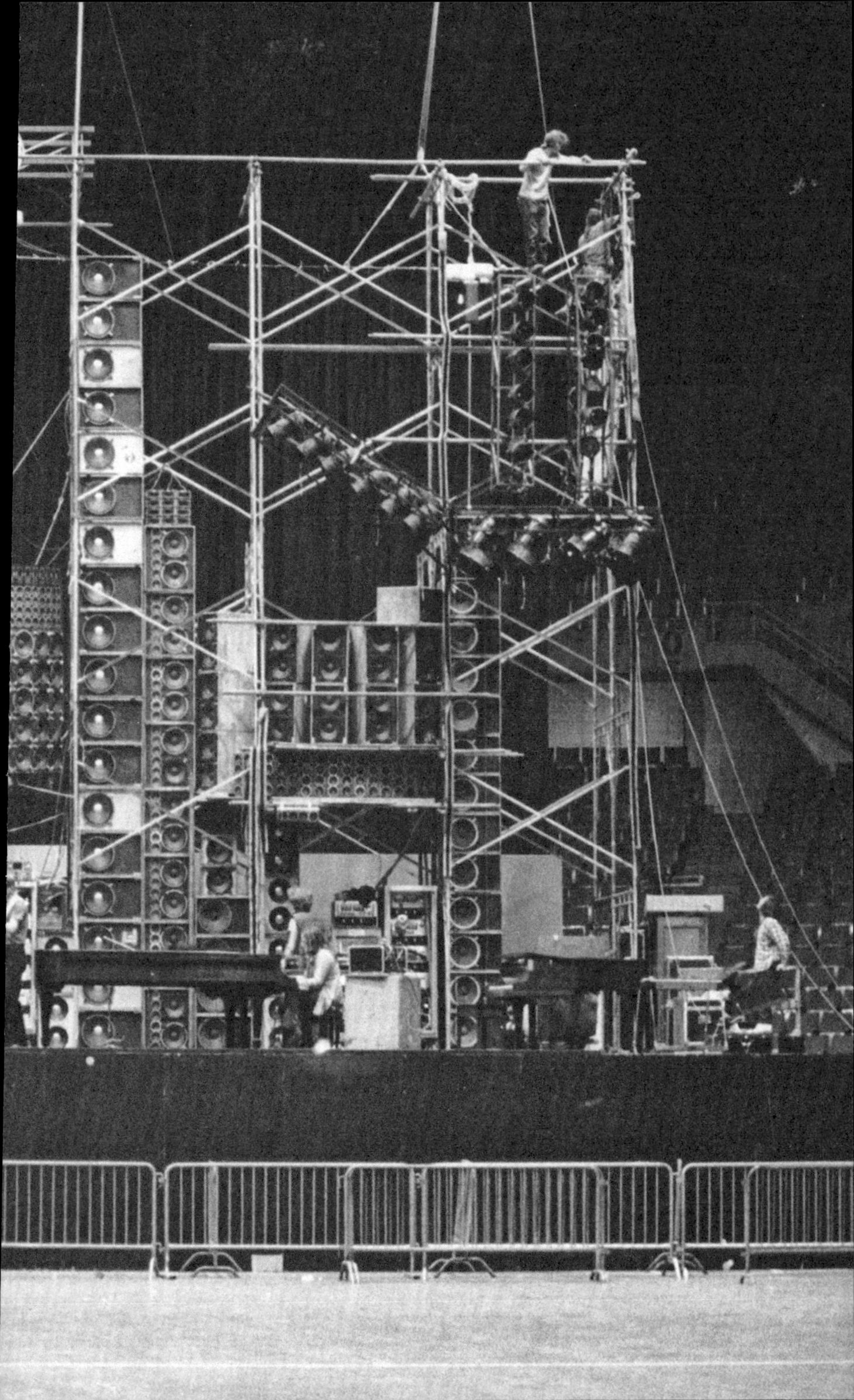

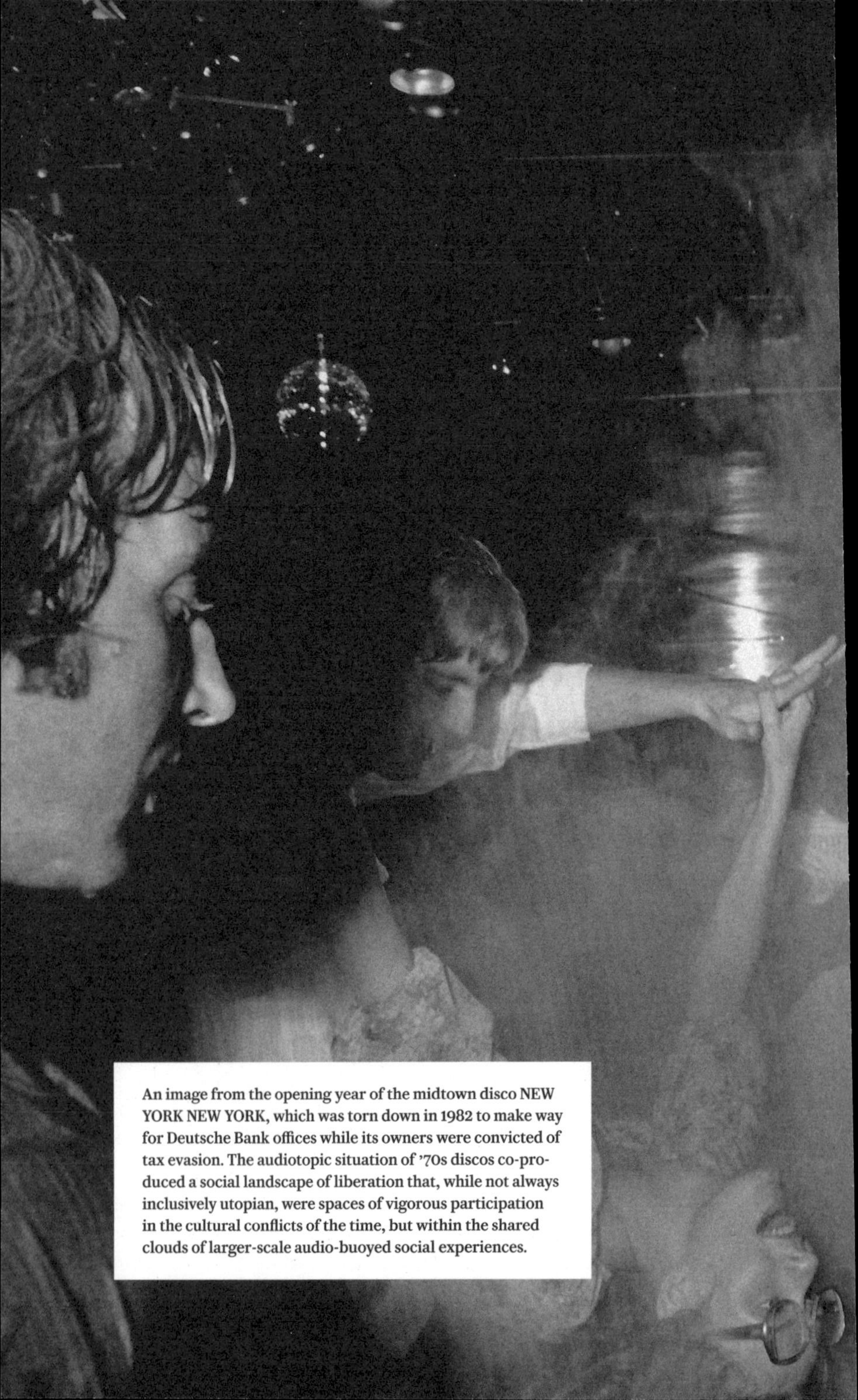

An image from the opening year of the midtown disco NEW YORK NEW YORK, which was torn down in 1982 to make way for Deutsche Bank offices while its owners were convicted of tax evasion. The audiotopic situation of '70s discos co-produced a social landscape of liberation that, while not always inclusively utopian, were spaces of vigorous participation in the cultural conflicts of the time, but within the shared clouds of larger-scale audio-buoyed social experiences.

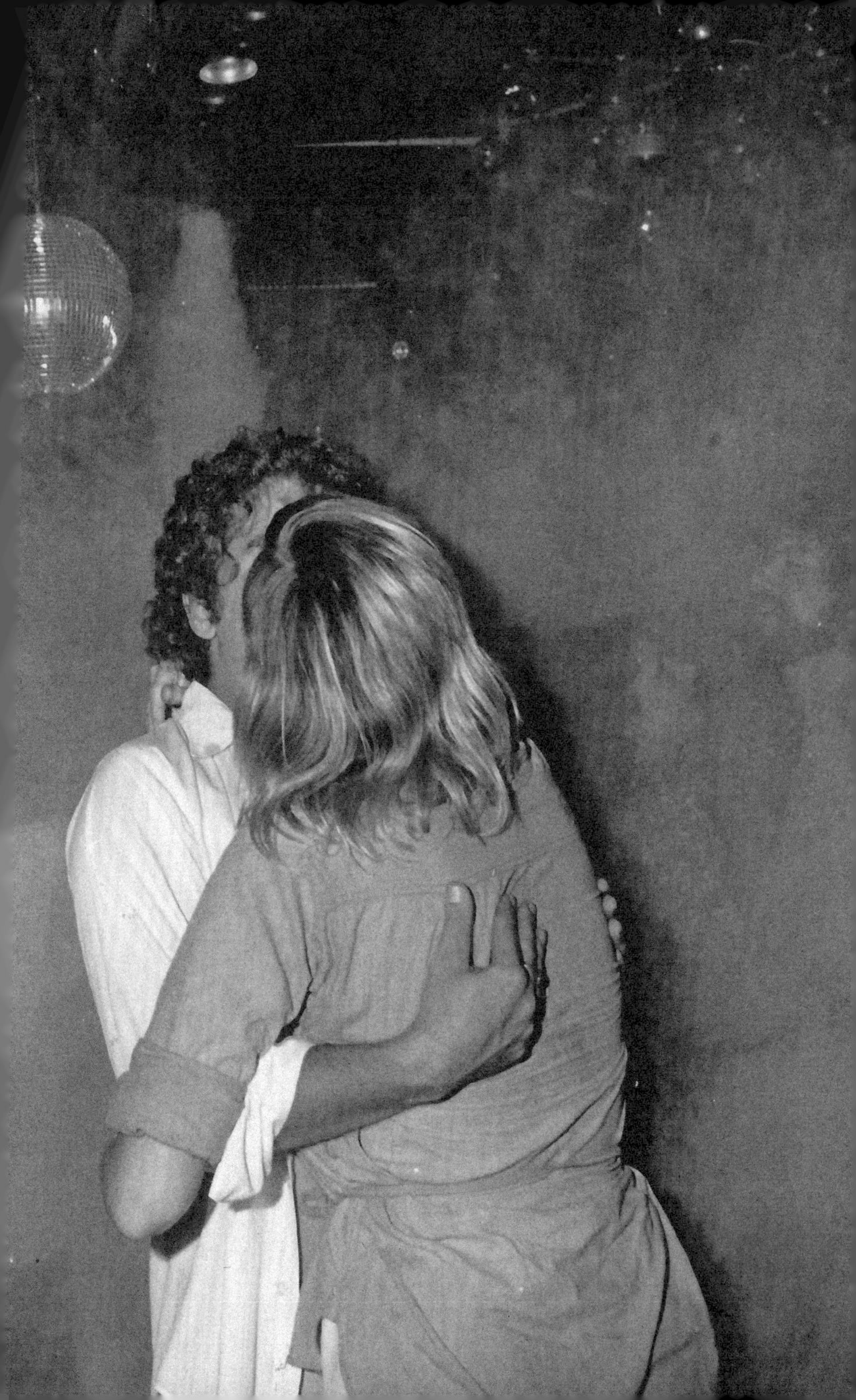

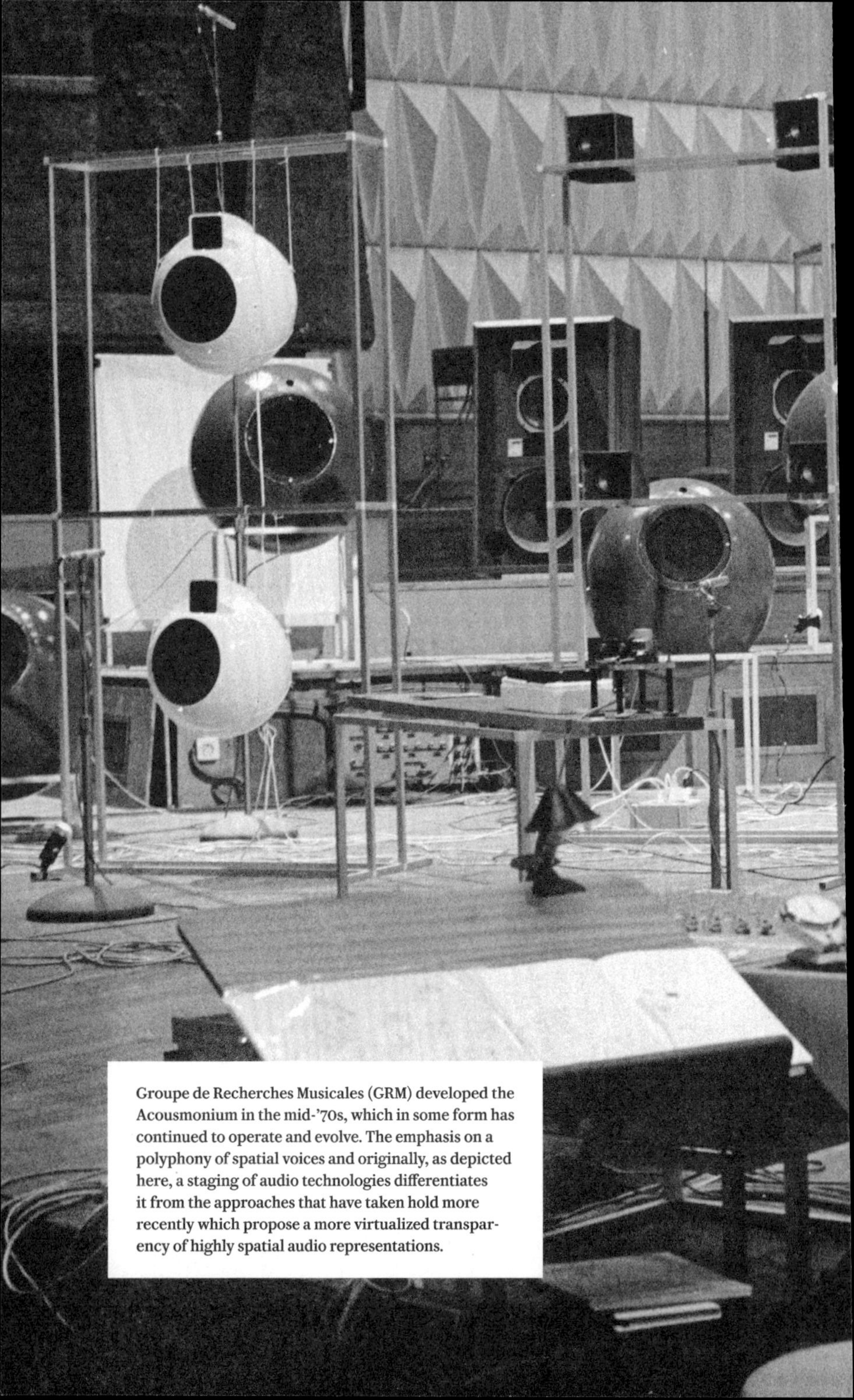

Groupe de Recherches Musicales (GRM) developed the Acousmonium in the mid-'70s, which in some form has continued to operate and evolve. The emphasis on a polyphony of spatial voices and originally, as depicted here, a staging of audio technologies differentiates it from the approaches that have taken hold more recently which propose a more virtualized transparency of highly spatial audio representations.

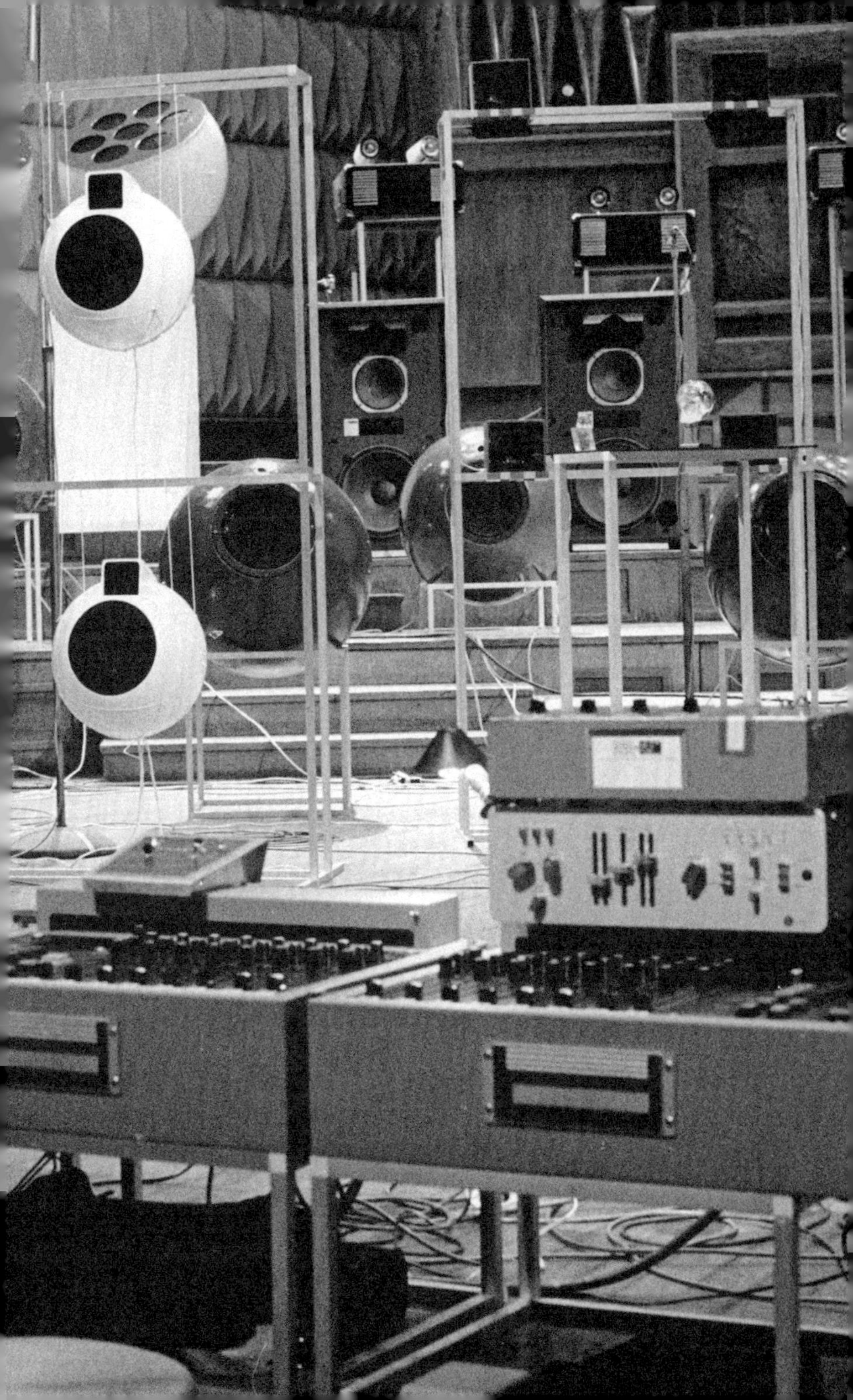

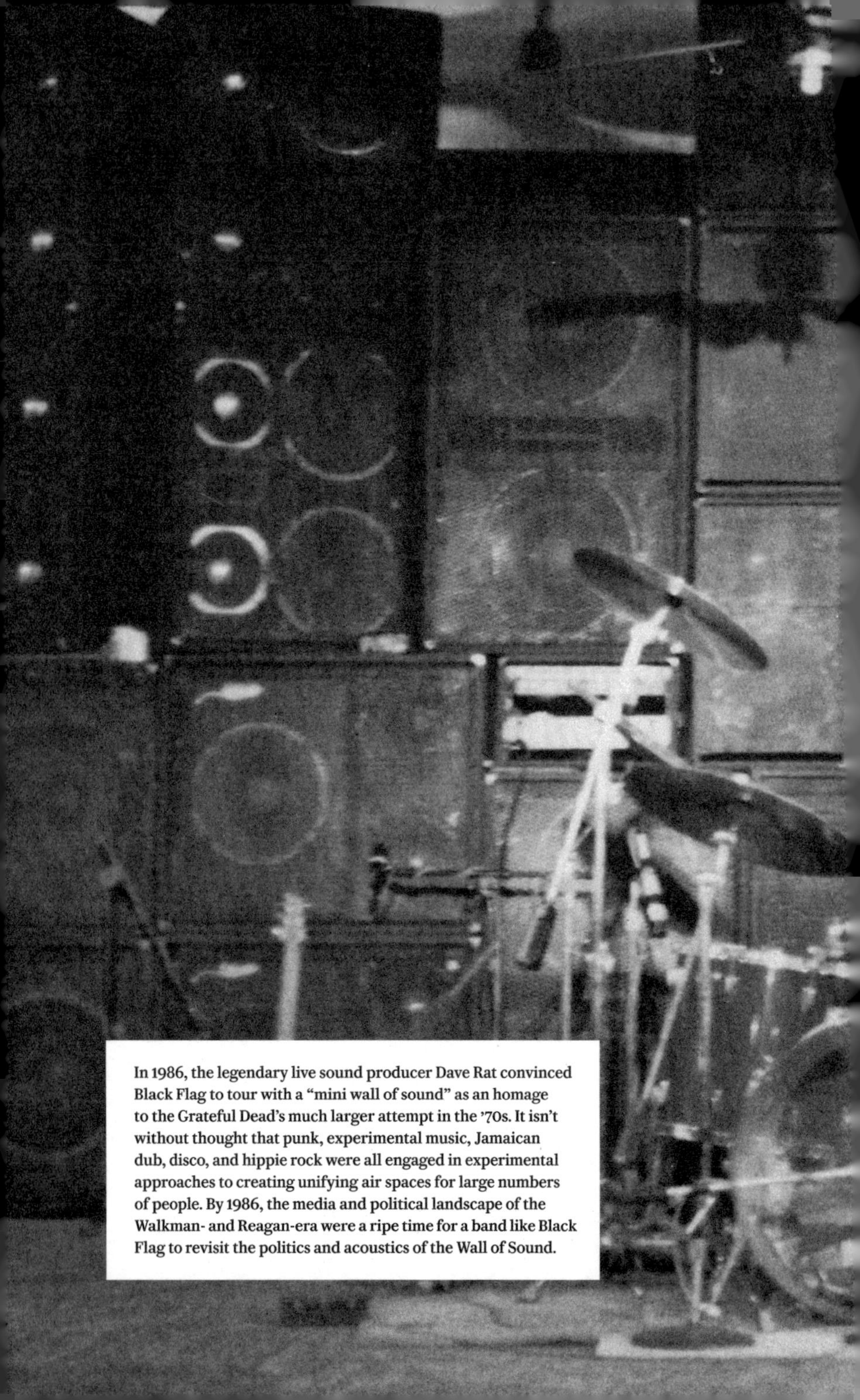

In 1986, the legendary live sound producer Dave Rat convinced Black Flag to tour with a "mini wall of sound" as an homage to the Grateful Dead's much larger attempt in the '70s. It isn't without thought that punk, experimental music, Jamaican dub, disco, and hippie rock were all engaged in experimental approaches to creating unifying air spaces for large numbers of people. By 1986, the media and political landscape of the Walkman- and Reagan-era were a ripe time for a band like Black Flag to revisit the politics and acoustics of the Wall of Sound.

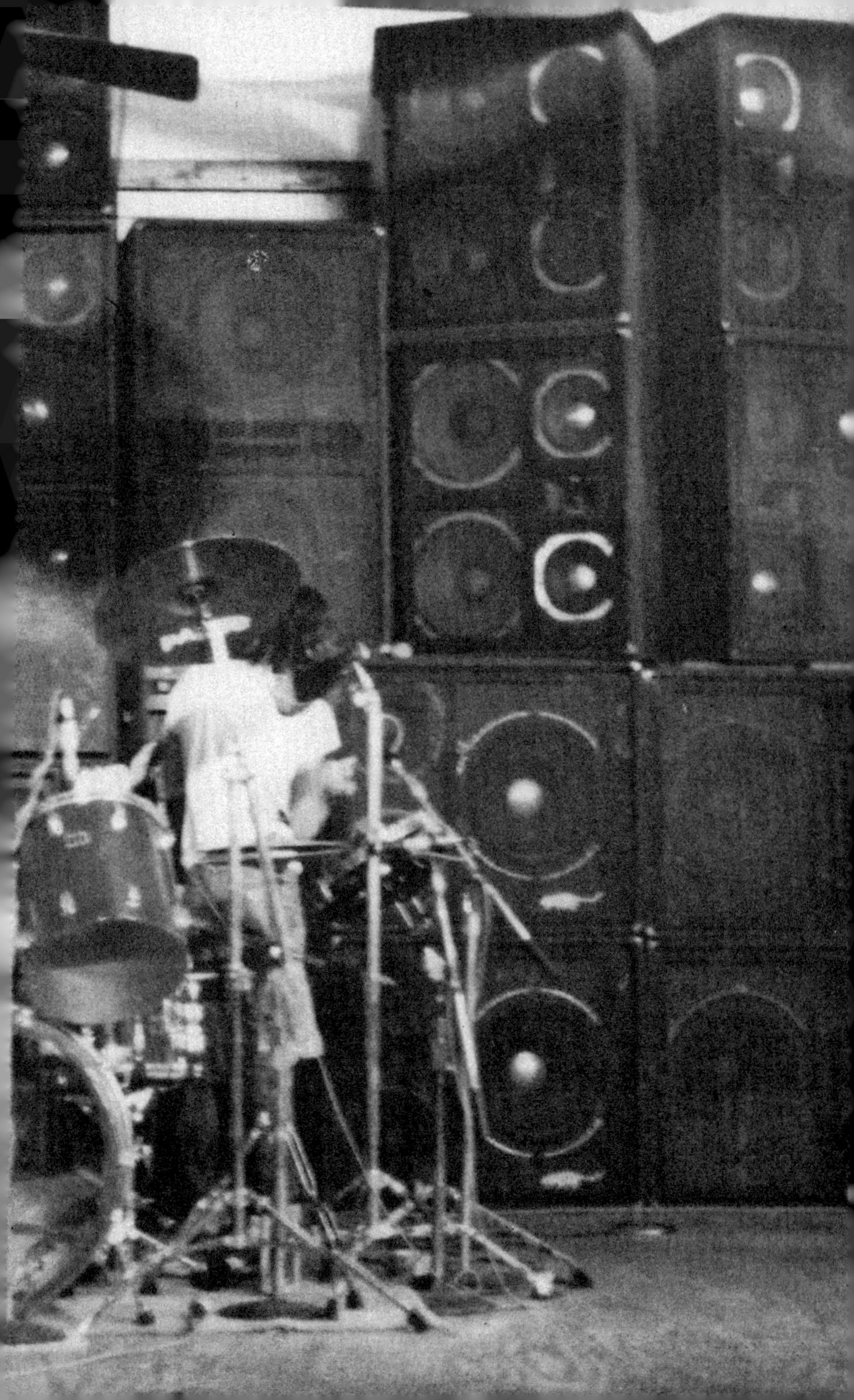

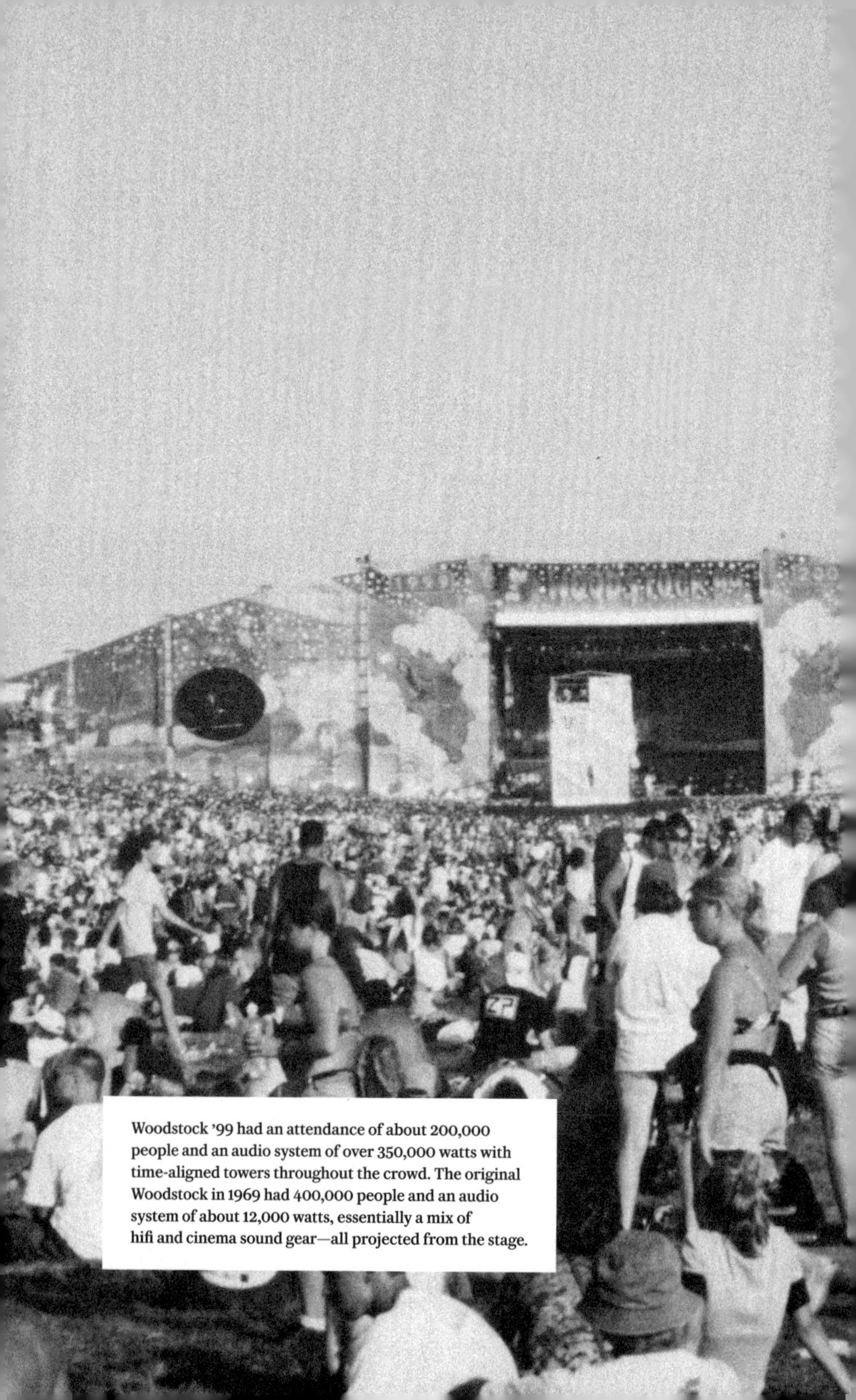

Woodstock ’99 had an attendance of about 200,000 people and an audio system of over 350,000 watts with time-aligned towers throughout the crowd. The original Woodstock in 1969 had 400,000 people and an audio system of about 12,000 watts, essentially a mix of hifi and cinema sound gear—all projected from the stage.

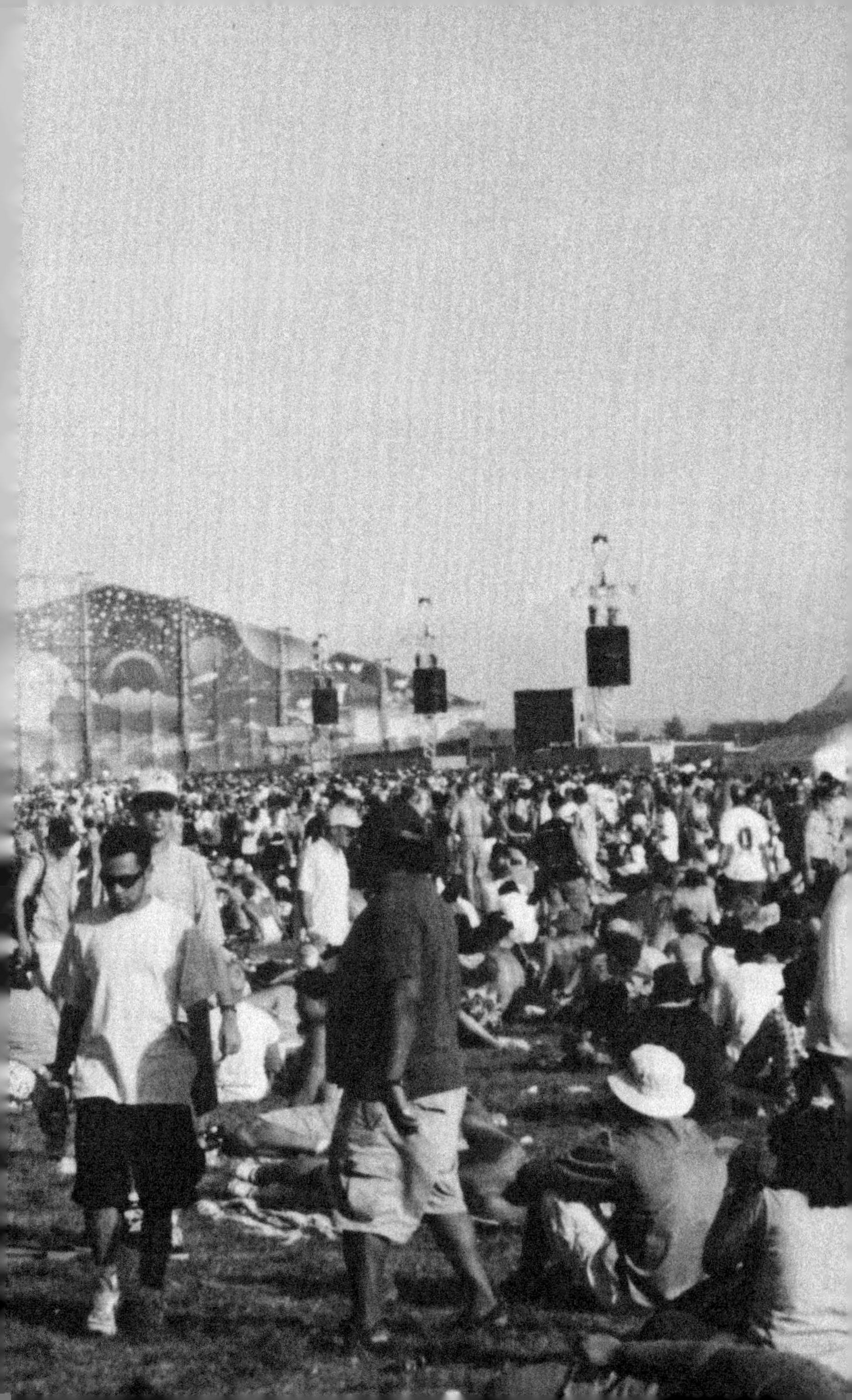

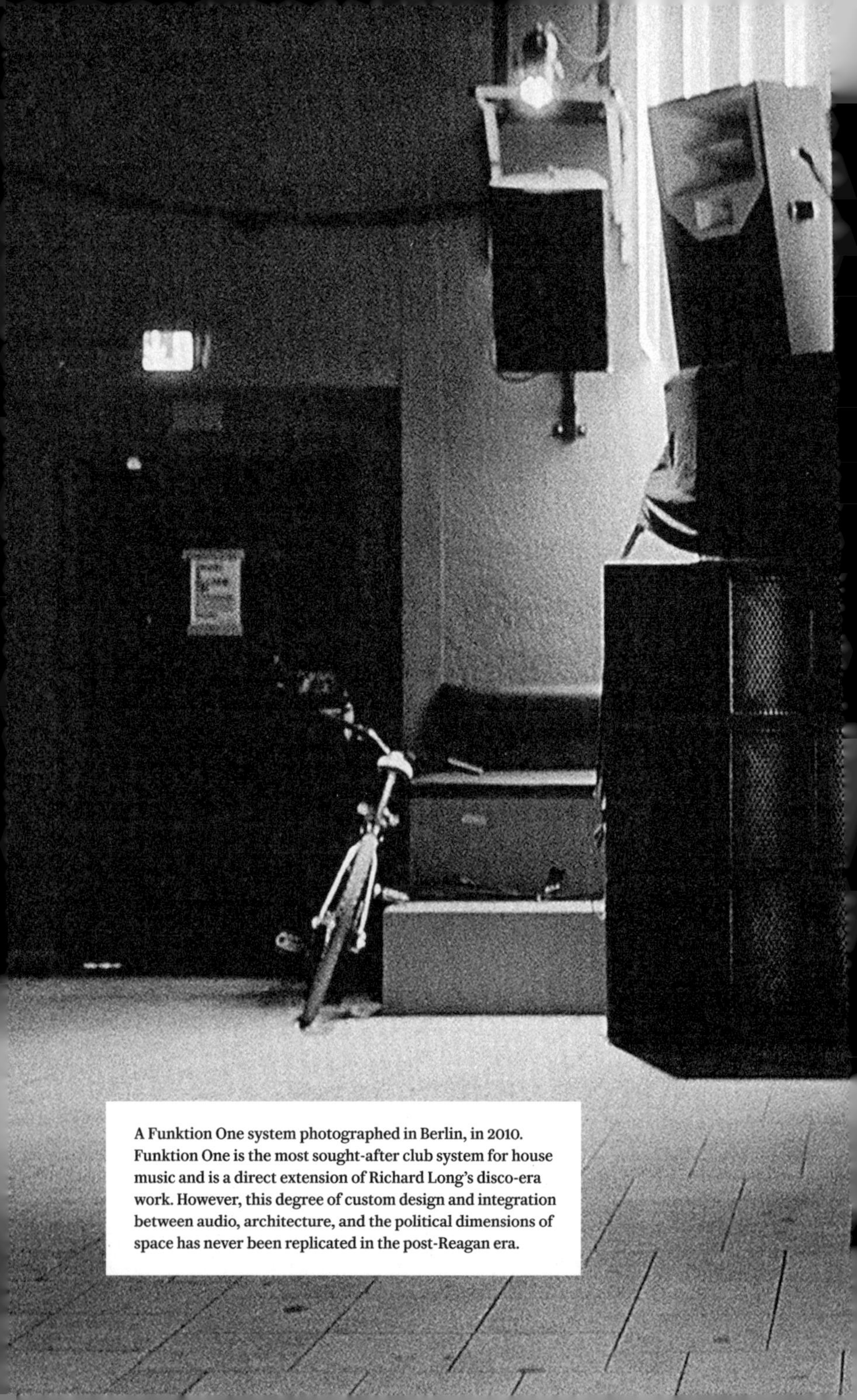

A Funktion One system photographed in Berlin, in 2010. Funktion One is the most sought-after club system for house music and is a direct extension of Richard Long's disco-era work. However, this degree of custom design and integration between audio, architecture, and the political dimensions of space has never been replicated in the post-Reagan era.

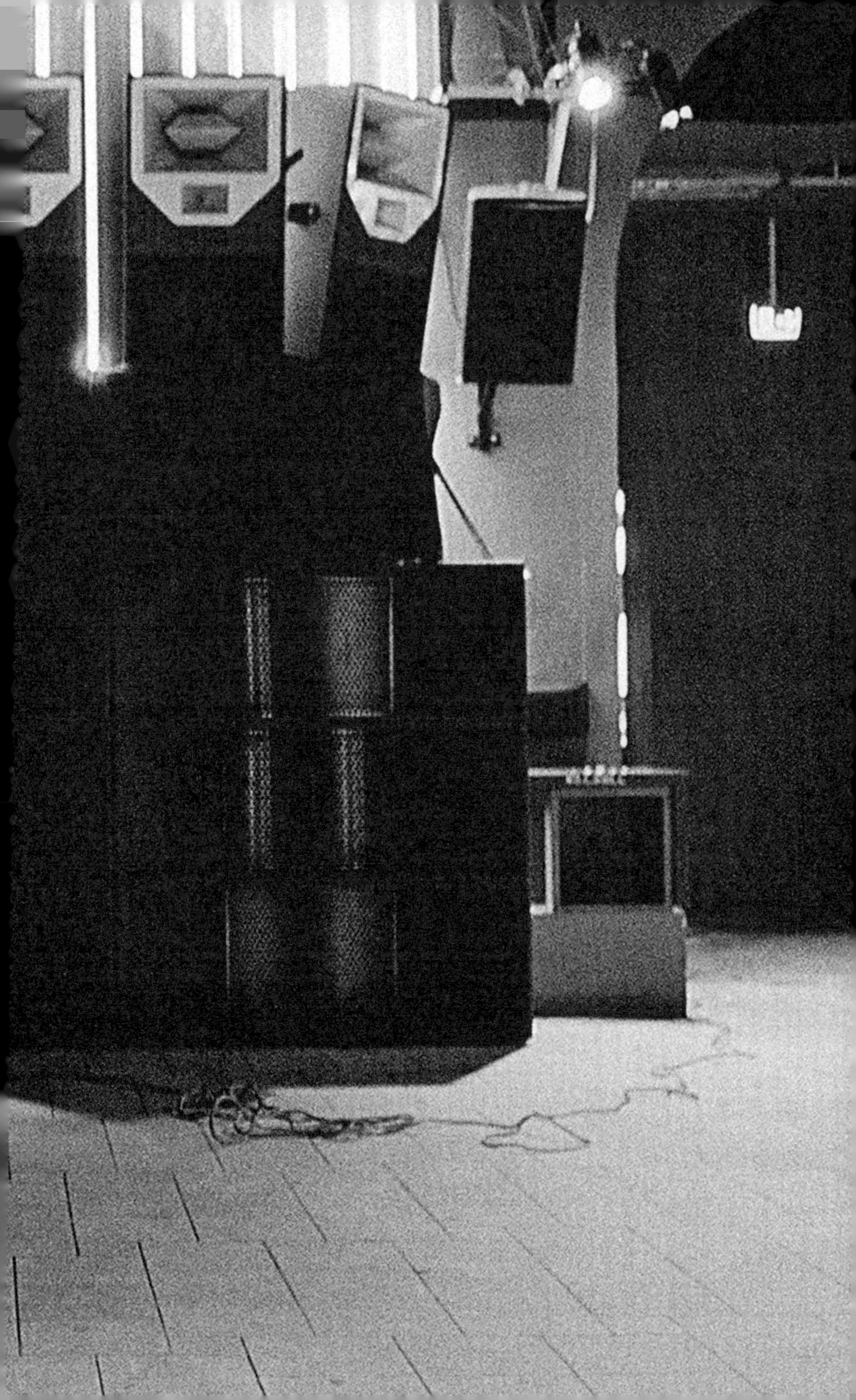

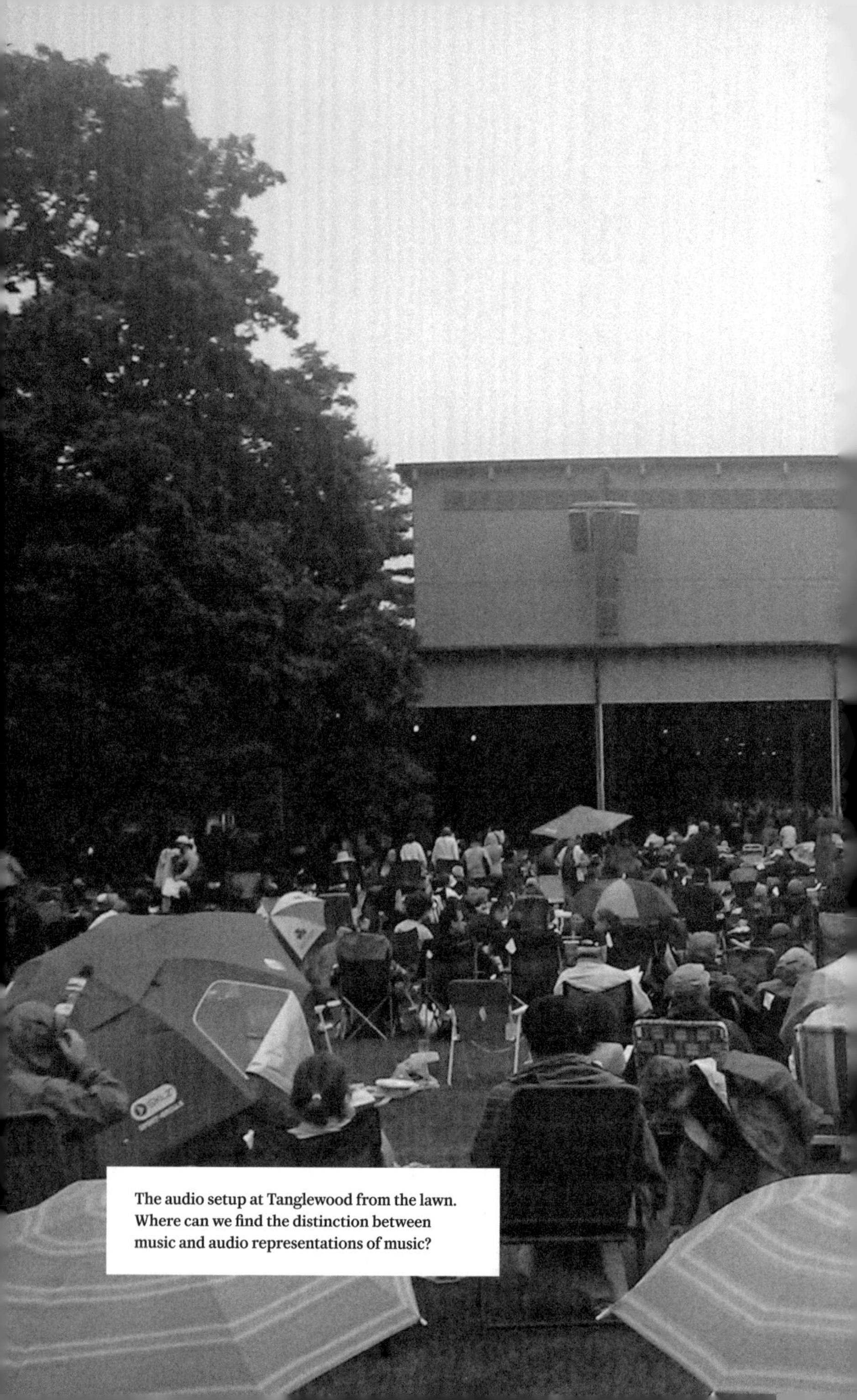

The audio setup at Tanglewood from the lawn. Where can we find the distinction between music and audio representations of music?

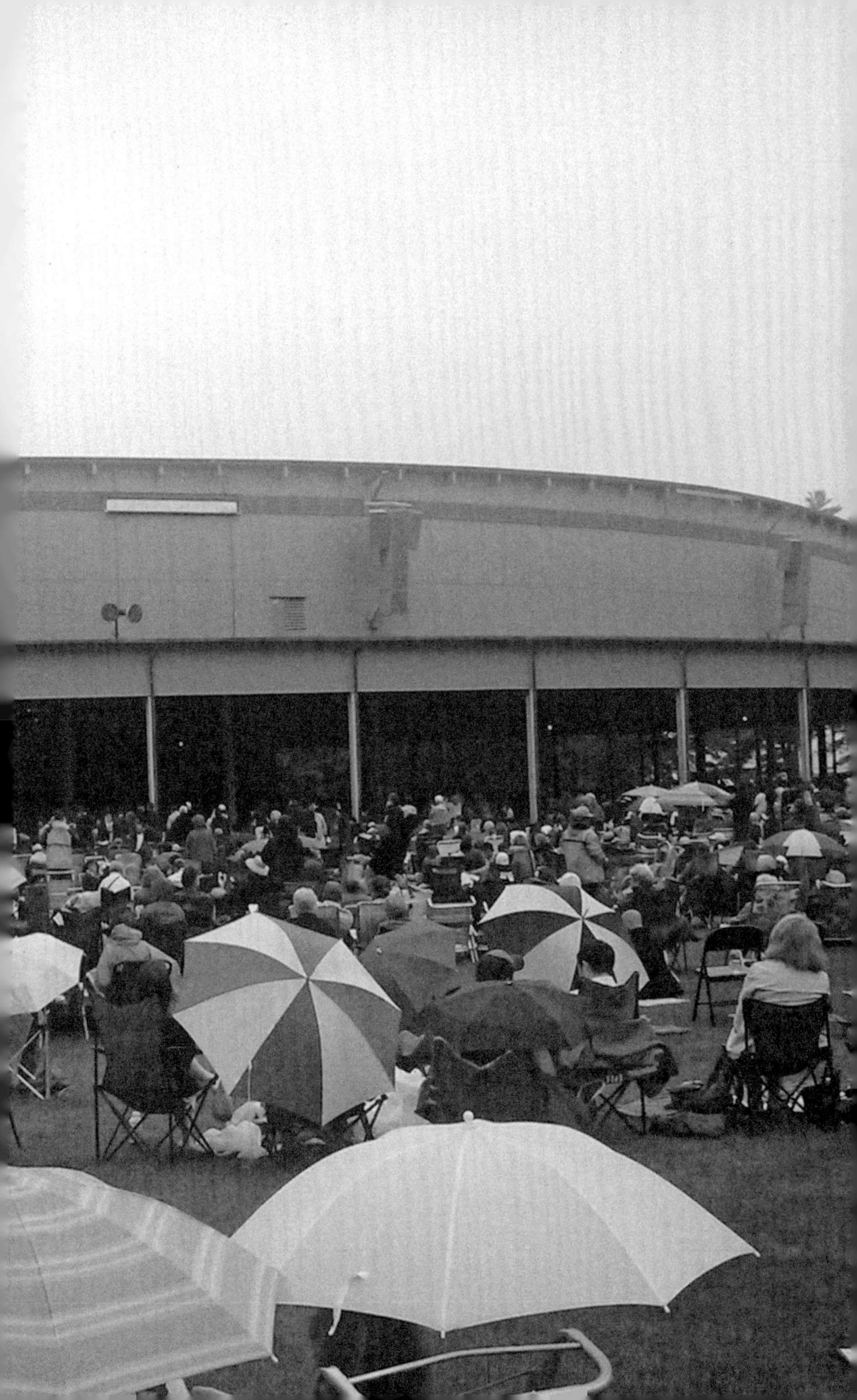

Figures in Air

Essays Toward a Philosophy of Audio

Micah Silver

INVENTORY PRESS

A proposed continuity

Radically Partial

These texts, written in 2012 and 2013, are for both a general audience of people tapping play on their iPhones, and for artists and musicians who engage with audio in their work. I hope I also speak to the curatorial field: those thinking through what it takes to adequately situate and position "sound art" which, as will be explained, is most often leveraging the potentials of audio more than sound. I've tried to write directly and with as little shorthand as possible—perhaps at the expense of flourish and depth. In doing this I expect experts will find some lack, but for anyone who is curiously toying with the idea of falling in love with audio and testing the resistance of it's materiality and social potentials, I believe the ideas in this book at the very least pose many useful questions.

My intention with these texts is not to stake any particular lasting claim, but simply to offer some possible systems of resonance; a humble contribution to understanding an aspect of our time through the lens of audio. These questions have been guiding my evolving participation in culture—as an escaped musician turned artist and as a sometimes curator. But mostly these essays are born of a person born in 1980 to a culture in which certain technologies of representation had already perforated our life-awareness completely (the Walkman Generation one might say, if playing names), and perhaps this text is an excavation of some of what we inherited, what we missed, and what we must decipher.

This book is about thinking clearly with audio, where we can locate it, and what it means to use it. The word "audio" here is not invoked as a semi-interchangeable conceptual unit with sound, music, or recordings, let alone sound art. As you'll come to discover, we're carving out a very particular, but then not particularly delimited, expanse here for the purposes of both consequence and possibility; a defining of terms for a relation to audio. In doing this, I hope to raise the disconnection between audio and music. These shortcircuits are important to note. The book is as much about music in absence as it is about audio articulated.

The intent here is, in part, to resituate the dialogue from the page, the stream, the encoding; from the listener, the type of listening, from the ear, from the mind, and indeed from the individual human; *to the air*. And in the air we find ourselves together, and in doing so we can account implicitly for the more conceptual and historic locations for audio's potency. Air is the material for the composition of a commons. Audio is an ephemeral social architecture made of air.

The story being told is narrated between conceptual proposals and a selective history of commonizing audio, of social desires driving technologists, of the communal prehistory of what audio's primary use-case has become post-Walkman (not to mention, post-Internet). It's for us, for finding new criteria, and for pushing away ambivalent pessimistic relativisms and inconsequential binaries. We're after audio's potential for social purpose.

For ongoing support at different stages of the book's development: Eugenia Bell, Azra Akšamija, Sherry Turkel, Johannes Goebel, Florian Hecker, Adam Michaels, Grace Robinson-Leo, Justin Luke, Anne Callahan, and the late Susan Sollins.

1 Listening Beyond Turing

The sense a listener may have, when listening to audio, that an originating moment—a duration—transfers into the present from another time is a function of our imagination, not a property of technology.

The ongoing universality of this fiction as indivisible from our listening experience isn't made more or less imaginary given the fidelity of the audio to the actual duration. We should consider instead whether the status of audio is even best understood as a reproduction or representation of this supposed origin. Why do we even believe that audio is an artifact that arrives to us from a past—and why do we trust that this past existed at all, regardless of whether it would be possible to represent it, never mind reproduce it? We use the words "recording" and "reproduction" with audio technologies, without having a clear idea of what exactly is recorded and could be reproduced. To excavate these questions in hopes of locating a more concrete and conscious relationship to these possibilities of understanding is the purpose of this essay.

• • •

As we listen to audio, we imagine an authorial human presence and deduce from its existence that what we are hearing is in some way, however small, a human testimony to the existence of an Earth and of a civilization that we know something about. We can believe that audio are telling us something about this civilization.[1] We can never

1 Audio will be treated as a countable plural noun. The book will expand on this use as it unfolds.

exclusively hear sound, as say, a microphone without a consciousness does. We always listen, and while listening is the resonant interface to the world via the ear, the ear does not simply hear, it also speaks—it represents—and is immediately subject to our mind: a system or non-system whose nature remains unresolved. Rational listening is synonymous and indistinguishable from the imaginary, and both are inseparable from memory.[2] An example (also an exception): we are holding an object whose function is a temporal displacement. Let's say it is a reel of magnetic tape or a shellac disc—as tangibly physical as audio has been, and no longer is. The metals of the tape were shaped just moments before along the path of our own voice, the disc cut into a topographically ornate spiral of valleys that trace the movement of our lips. But if we aren't holding our voice in our hands, contained by these materials, what is it that we hold? To understand audio, we need to take great care in unraveling what, if anything, of our voice, we can believe we are holding, and gather the plurality of audio's imaginary so as not to confuse the way we can know, or can hold, a voice.

Another modulation: as we listen online to a "real-time" feed of a microphone placed deep underwater in the oceanic waveguide, or in a department store, at a live concert, or from a webcam in someone's garage: in these cases, are we to believe that nothing originary is transferred across time and space? What do we believe we know about these places by listening via the remote microscopy of a microphone? And maybe more important, why do we desire to believe these things against so many factors that would dissuade us if we compared these beliefs to our knowledge of the mediations present?

In the first case, we could not be there listening in the same way—with scuba gear we would hear our breathing, or in a submarine we would hear the machine and listen through another microphone, but be there in time. For the others, what is the difference in how we imagine listening as we read, in this moment, either by being there or by being elsewhere, imagining this other "there." Our experience right now, projecting these examples, can tell us a lot about these differences, and in many

2 "Speculative Solution: Quentin Meillassoux and Florian Hecker Talk Hyperchaos," in *Urbanomic*, July 22, 2010.

ways, text—when used as in this paragraph—can be quite close to the procedures of audio within our imagination. What does the author's voice sound like? Or is it *your* voice, using the author's words to speak silently to yourself?

Yet the misperception in developing a resolution to this possible similitude of originary and representation is that we can never be in both locations at once to compare. The differences and the slope of these discontinuities in meaning with the control of our own subjectivity (however contingent) are not possible. No human has experienced this. But audio's intersubjective basis, our implicit social contract around what it means, is clear: we believe that there does exist some sense of testimony in the experience of audio that transfers not the entirety of the originating duration's qualia (or even an acoustic reality of it), but just *something* of it. This something is enough for most of us to have an active and ongoing relationship to audio that affords it a knowledge status that we would never ascribe to a photographic image, a drawing, or even a video. Imageless, we imagine more. Listening, we are vulnerable to suggestion, hypnosis. Audio is a kind of transport, but from where to where and how?

• • •

Much has already been written about the history of audio from the perspective of recording and playback technology, notably by Jonathan Sterne in his 2003 book *The Audible Past*, the release of which established "Sound Studies" as a possible inter-disciplinary field between cultural theory, history, and musicology. As illustrated in his text, the early promotional strategy for sound recording and reproduction was the claim of preserving the voice beyond death, and in the intrigue of being a listener to the deceased.[3] To demystify audio as an idea, different from the technology of sound recording, we can turn to Sterne and see another nineteenth-century innovation as an antecedent: embalming.[4]

see page 1

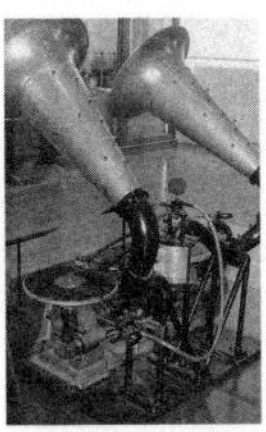

3 Sterne, Jonathan, *The Audible Past: Cultural Origins of Sound Reproduction* (Durham, NC: Duke University Press, 2003), 289.

4 Ibid., 293.

Audio as in the preservation of a body to be witnessed post-mortem, transforms the loss of the essential features of "being alive" into the representation of life. Why such a necrophiliac desire for memory in the absence of a memory's real? To view a "designed" or embalmed corpse is to construct an artifact for our imagination and memory to enhance. We know the corpse is dead and we mourn it and feel it is a part of a different non-life as we perceive it, but nevertheless it affords us a momentary option for a personal transubstantiation of the corpse into imagined life.[5]

Given that audio originates and develops technologically until the mid-twentieth century within explicitly Christian societies, the connection to the Eucharist—the transformation of bread and wine into the body of Christ—is worth a close examination (and in fact our relationship to audio is more similar to this than our relationship to the embalmed loved one). To go further with this religious framework for audio is beyond the scope of what this essay will attempt, but what *is* important to consider is that "audio" as an idea existed as a schema and a desire long before the technology embodied the desire. The origins of this desire can likely be traced to a Judeo-Christian foundation.

Regardless of how or why the seed was planted, audio are uniquely tied up with our sun-staring relationship with death. The arrival of the post-death representation via sound and the sensuality of its unquestionably present and real acoustic touch short-circuit our criticality. It takes shape for us with imagination, with fantasy, and with belief. Increasingly, some relate scientifically, technologically (a demand for a specific encoding, a format preference, a listening situation), but how can we relate to audio as a vehicle for observing our relationship to its potentiality, not dismissing the fantastic, the artifactual, the critical, the syntactical, the sensual? This has proved difficult for the culture to develop.

• • •

Now, back to the tape we hold in our hands containing a *something* of our

5 The earliest remaining recording of Edison's voice is from a series of wax cylinders on which he narrates an imaginary journey titled "Around the World on a Phonograph." This 1888 recording evidences Edison's awareness that imagining places via audio was an important aspect of his technology's potential. Part of the recording can be heard online: http://archive.org/details/aroundworldonphon1888.

voice. We can likely agree that the duration[6] we experienced while making the recording has entirely disappeared. This duration is what we felt, how long we felt it existed, the informational crosstalk of our other senses that influenced our time sense during the recording, the other presences in the room, the moments that preceded its effect. This aggregate temporality sensed in the moments passing and remembered as an episode is what is meant by duration. This experiential unit has nothing but a correlational, linguistic relationship with the quantization of episodes by a clock of any system, not only philosophically, but also practically.

Human beings are terrible at relating the two without the ongoing reference to a quantitative device.[7] A duration for us need not be limited to the clearest case of recording in time; it is also the shaping of sound into a duration *by* a human being that accomplishes the same thing. *So the audio production studio, in any form, is a means through which durations are composed.* Temporalities are constructed over many episodes, aggregated into a composite temporality, objectified so it can then be represented in the future via playback.

So as we listen now to our tape and observe the relationship between our memory of the original duration recorded and our unclocked sense of it via the audio representation, we are left with a somehow related, but additional impression of the duration. This is an overlay in the most lucid moment, but quite likely is cause for what Jean Piaget called a "deforming assimilation," which we produce via an egocentric desire to relate the sameness of our experiences rather than to understand deeply the difference and construct new schemas for what is possible based on subtle interference.

We are mutating our memory in this act of listening. And as we listen, likely we remember not only the duration of the inscription as an overlay, but other overlays as well. Here we can turn to philosopher Henri Bergson, theorist of duration-based theories of time and

6 Bergson's 1896 publication *Matter and Memory* emphasizes a particular definition of the word "duration" which is used to explore in detail the relationship between sound, listening, memory, and language—all components of phonographic experience's cultural transforms. See Bergson Henri, *Matter and Memory* (New York: Cosimo Classics, 2007).

7 Brown, Scott W., and D. Alan Stubbs. "Attention and Interference in Prospective and Retrospective Timing," in *Perception* 21, no. 4 (1992), 545–57.

memory, and consider how little choice and control we have in which memories we recall and for what purpose our mind identifies them as salient information for informing the present.[8] We aren't wonderful at controlling what we remember when we listen, where our senses go. Most of us can barely meditate for a minute holding a single word or thought without interruption of some memory, anxiety, or curiosity. With audio we are nearly hypnotized, open to suggestion.

So during playback, this sensual crosstalk is a new constellation; we can't remove it from our experience. Even in an anechoic chamber there is no escaping the aggregate sensation of *being where we are* to some extent. In this way we can understand Cage's famous observations about listening within such a space to be profoundly myopic in their emphasis on the independence of aural listening and for leveraging a notion of the aural that lacks intersensory contingency. One can also wear extremely powerful earmuffs or earplugs and hear the flowing of one's blood and the conduction of one's bones, anywhere. Silence on a rollercoaster has potentials unique from silence in an anechoic chamber. There is not one arrival or charge to silence, there are an infinite number with distinct possibilities. But nevertheless, if the tape machine was calibrated accurately, we know that a quantification of the originary duration is being represented during playback with as much accuracy as was possible during the inscription. That quantity is placed into a new time, making a new episode of durational experience, and has, regardless of context, a spell-like quality that can be distinguished as a constant among other perceptions.

• • •

Temporaneity

This quantified fragment of a temporality objectified (an objectification of memory) is the conduit between the originary moment and the present. We'll call this *temporaneity*.

The other information that flows with audio as an artifact is vast and this definition in no way reduces the

8 Bergson 2007 (note 6), 133–57.

density and diversity of those components. But, these are contingent on the durational relocation described. Without this, the other components become ungrounded and singular representations of their own properties with the mythology of audio—an originary duration relocated, redacted.

We can prove this to ourselves in a number of ways, depending on the technology. With digital audio, all we need to do is iteratively degrade a recording and observe what is left as we listen to a two-bit representation and compare it to a sixteen-bit representation. Timbre and the perception of layers or depth of sound disappear as the possible dynamic range within the representation are squashed and become distorted by increasing noise. What remains most clearly is a beginning, a temporal gesture marked by sound, and an end. Sixteen-bit audio has 65,257 possible values, whereas a two-bit representation would have four. Yet if we compare this two-bit representation of our voice to a two-bit representation of a piece of music we know, the existence of a difference between the two will be clear. In each there remains a trace of the origins' temporality, however mutated.

If we perform the same experiment in the time (and collaterally, frequency) domain and degrade the sampling rate (the number of temporal instances or slices of data extracted from a source within given clock duration) from the common 44,100 samples per second to 1000 samples per second, we reduce the possible frequency range from the full range of human hearing to a maximum frequency of 454hz (just above the tuning note, A440 within the range of most instruments and the voice). If we go further and lower the sample rate to below twenty samples per second, we would begin to hear not a continuity of sound, but a stuttering pulse of sound with changing spectra limited to 44hz, an extremely low "bass" sound. In this stuttering, the representation of a duration is replaced by a new, rhythmic foundation which is heard as its own duration, even though all sampling rates are functioning the same way. This threshold between the functional illusion of a non-quantized representation and a quantized one is the critical Turingesque test of audio: *it is human testimony if it is a continuity*. Even with a lack of spectral realism, we will still attempt to compensate with our imagination toward belief, especially if a voice.

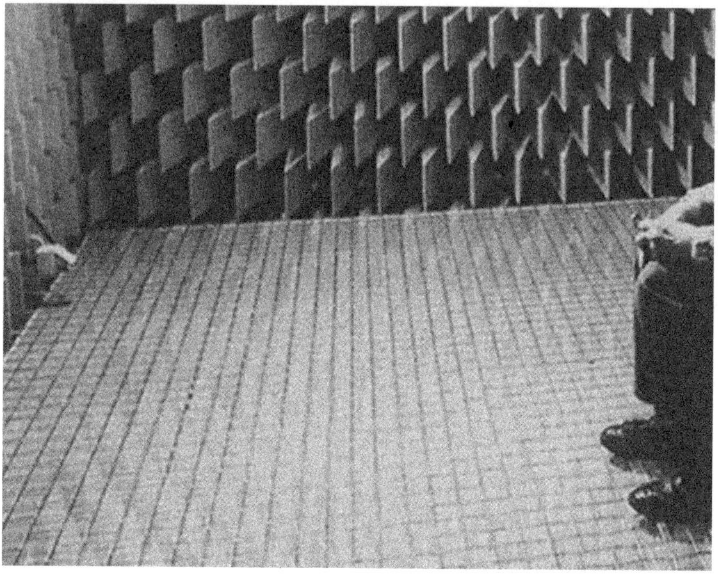

While the air in an anechoic chamber may be notably without compression waves, that is, sound, any notion of silence that assumes perception can be parsed sense by sense misses the fact that even shoes are impacting the flow of blood and the position of one's hands—the nervous system's observable state. Perhaps a more radical position than imagining that there is no silence is to say that there is no listening with only ears.

Technologically we achieved this Turing-passing moment almost immediately with a phonograph's method of spinning an inscribed circular object against a moving blade. Due to this lucky history of invention, human beings were not afforded the same critical process we've engaged with around artificial intelligence and robotics, for example, and so the necrophilia and imaginary, the sense of obvious realness that audio brings, has gone unconsidered or at least under-theorized. Audio's utility is like music, treated as given, it just seems to be as we use it. We provide the testimony to ourselves that there was an originary moment at all. This testimony holds even if the representation of this past to our senses, played back in a new time and place is decidedly subject to mathematical transformation whose details either cross or don't cross important thresholds of believability.

The augmentations we bring via memory and imagination to this experience of sound, to our translated duration depends on this *belief* that another time is being represented. Yet this testimony is not required for us to sense that there is a transmission across time; all that is required is a technical threshold that audio technology has never failed to achieve.[9]

9 There have been technological precursors to audio that in retrospect we can understand as "audio technology," but at the time they were understood primarily as tools for automatic writing. See Sterne 2003 (note 3) for a thorough overview. Note that it is assumed here that timbre carries tremendous amounts of potential meaning for a listener and that "duration" is an underlay for this information.

2
Audio's Ancestrality

Understanding time as divisible into quanta is the conceptual and technical precursor catalyst for audio. And while we think of this today as a feature of digital technology, to imagine it limited to these contemporary resources obscures the full weight of this rupture.

The ecological historian Alfred Crosby argues that the rapid transformation of European society from a muddy, diseased mess into the global superpower it became occurred between 1250 and 1600. He attributes this transition not to technological development, but to the rapid adoption of quantification in domains governing our understanding of time and space. Europe would be the first region of the world to achieve "pantometry" or universal measurement and the embodiment of these new ways of thinking afforded the paradigm shifts to follow across the senses and their disciplinary bodies.

The Islamic world and Asia had long preceded Europe in inventing systems of measure, but had belief systems that weren't adaptable through the "glare of clarification" present in an evolving Europe.[10] For example, the first clocks in Europe were not constructed until c. 1270, while water clocks in various forms had existed in ancient Egypt and developed by several polymaths in what is now the region between present-day Syria, Iraq, and Turkey. Al-Jazari, for example, published *The Book of Knowledge of Ingenious Mechanical Devices*[11] in c. 1206, which

10 Crosby, Alfred W. *The Measure of Reality: Quantification and Western Society, 1250–1600* (Cambridge, UK: Cambridge University Press, 1997), 57.

11 Ibn al-Razzaz al Jazari, *The Book of Knowledge of Ingenious Mechanical Devices: Kitáb Fí Ma'rifat Al-hiyal Al-handasiyya* (Berlin: Springer, 1973).

included numerous mechanical devices that could execute consistent periodicities, including clocks and many other inventions. But even preceding the clock is the notion of counting reliably. The abacus was in use in China between 500 and 1000, a period during which Europeans had no addition or subtraction symbol and were using "finger reckoning," which was never standardized nor capable of managing large sums.[12]

So what was the engine that drove a desire for quantification through music notation, perspective in painting and drawing, introduced double-entry bookkeeping, and moved pre-science toward science?

Of course there is a multitude of forces, but one is most intriguing for the dialog that follows: a reciprocal dynamic between the populist, common sense understanding of the world and scientific and pre-scientific thinking among specialists increasingly sanctioned by the governing power. This exchange would shatter what Crosby calls "The Venerable Model," a contained view of the world that met the needs of the population's comfort zone for reality's functionality, with ideas about time and space that were verified as more true, yet were beyond common sense or empirical investigation.

see page 2

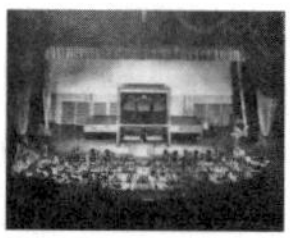

One lucid example of this is the Copernican Revolution sparked by Copernicus' 1543 text *De revolutionibus orbium coelestium*, which informed the public that their planet (and according to Kant, their subjectivity) was *not* the center of the universe. The text contains tables and diagrams incomprehensible to any non-specialist without expert knowledge of astronomy and mathematics, two fields few would have had any familiarity with. Yet it took over two hundred years for his model of the world to overwhelm the more commonsense view that, of course, we (or the "I") are the center, since pending an ongoing transcendent state there is very little we can experience with our senses in the absence of science to prove otherwise.

Thomas Kuhn's text on Copernicus suggests that the reason it took two centuries was not because the science was in question, it was due to the tensions and resistance to the shifting role of subjectivity in relationship to how knowledge is understood as such.

12 Crosby 1997.

Copernicus' "proof" was of a knowledge form unverifiable by the population, which aside from direct transmissions from God or an oracular divination at Delphi, was an emergent phenomena.[13]

Ancestrality

Quentin Meillassoux's term "ancestrality" can help to understand the cultural position of audio, a "discourse that includes a temporal discrepancy between thinking and being,"[14] and relates to the Copernican Revolution via audio's capacity to translate a beyond-death imaginary. This feature of audio has existed from its arrival and with technical means that for the general population remain empirically beyond grasp and conceptually distant. Few can describe in detail the encoding and decoding of tape, the particularities of a laser "reading" metal embedded in plastic (a compact disc) or the signal processing of the simplest car amplifier.

Ancestrality for Meillassoux is: "any reality anterior to the emergence of the human species—or even anterior to every recognized form of life on earth."[15] Here he is most specifically talking about events such as the Big Bang, which we have no way of understanding without the precursor methodology and rationalizations of science. As his argument unfolds, this definition is expanded to include all discourses with a temporal discrepancy, which he refers to as containing "dia-chronicity."[16]

The products of these realities, once arrived to our awareness via the universality of scientific agreement are "archefossils."

This diachronicity is one of audio's primary existential confounds: audio are representations of a quantified versioning of duration, transubstantiated and projected as temporal overlays into the present. When experienced they produce a discrepancy between our belief in its origination from the past (verified by our knowledge that the media it is carried on is only permeable through the action of human beings) and its potentially drastic impact on our sense of being in the present.

13 Kuhn, Thomas S., *The Copernican Revolution: Planetary Astronomy in the Development of Western Thought* (Cambridge, MA: Harvard University Press, 1957).

14 Meillassoux 2010, 112.

15 Ibid., 21.

16 Ibid., 10.

Meillassoux suggests a sleight of hand we play on ourselves to enable a lack of criticality when confronted with an archefossil such as audio, what he calls "the codicil of modernity"—the codicil through which the modern philosopher refrains (or at least thinks she does) from intervening in the content of science, while preserving a regime of meaning external to and more originary than that of science.[17]

For non-scientists this signals a retreat from objectivity and a defenseless relationship between our subjectivity and a reality filled with archefossils whose proofs we don't understand. Since modernity, science has had a privileged, monopolistic relationship to non-religious claims (but with the unquestionable authority of religion) to describing properties of objects that do not require a constituting subject. The codicil is the mechanism by which he claims we (non-scientist/thinkers) have removed ourselves from the discourse of the absolute for fear of being understood as theological or otherwise unreasonable, irrational, and driven blindly by belief rather than thought. This avoidance leaves the identification and quantification of primary qualities' meaning, as well as any possible notion of objectivity to the methodologies of scientific verification. And for Meillassoux, science "does not experiment with a view to validating the universality of its experiments; it carries out repeatable experiments with a view to external referents which endow these experiments with meaning."[18]

What Meillassoux claims we fear is the accusation that we are simple dogmatist, essentialists, espousing a tightly wound phenomenology or dabbling in ambivalent relativisms. What intellectual move can we make besides acquiescing to the social contract around science's authority over the objective? Regardless, we can also count on science understanding non-scientific observations as naive in the best case, and heretical at worst.

In order to overwhelm the codicil, Meillassoux rehabilitates the notion of primary and secondary qualities.

We shall therefore maintain the following: all those aspects of the object that can be formulated in mathematical terms can be meaningfully conceived as properties of the

17 Meillassoux, Quentin, Alain Badiou, and Ray Brassier, *After Finitude: An Essay on the Necessity of Contingency* (New York: Continuum, 2010), 13.

18 Ibid., 17.

> **object in itself. All those aspects of the object that can give rise to a mathematical thought (to a formula or to digitalization) rather than to a perception or sensation, can be meaningfully turned into properties of the thing not only as it is with me, but also as it is without me.**[19]

Meillassoux is not claiming that a "primary quality" is something in dialogue with a universal truth, nor is it a universal truth about the object, but rather a property that a human being might feel inspired toward within certain epistemological realms and those that can be described with mathematics can be considered representations of in-itself qualities of the thing. His larger project, in part, is to free thinkers outside of the sciences to engage the pursuit of these qualities with an eye toward utility in the domain of meaning.

The sciences, in contrast, evolve paradigmatically.[20] Participants' ability to quantify reproducible experimentation form the basis of knowledge production, independent of an exploration of what the experiments mean, except as alibi for their instrumentalization. The sciences are also now the instrument of politics, and specifically, of politics that are increasingly inseparable from religion, compounding the authority over an open-ended subjectivity for those outside, thinking differently.

see page 28

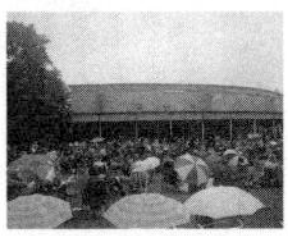

With Meillassoux's metastasized reclamation of primary qualities, we can embrace with clarity the opportunity to explore what "sound-in-itself" might constitute if not dismissed as a vague and intoxicated "Husserlian essentialism."[21]

If we can do this, perhaps we can illuminate the poly-dimensionality of audio's thingness and avoid adding further dichotomies and argumentation to a discourse that might ultimately be more about taste than important distinctions within reason, imagination, both or other.[22]

19 Ibid., 10.

20 Kuhn, Thomas S., *The Structure of Scientific Revolutions* (Chicago: University of Chicago Press, 1996).

21 Kim-Cohen, Seth, *In the Blink of an Ear: Toward a Non-Cochlear Sonic Art* (New York: Continuum, 2009), 178.

22 Boulez, Pierre, "Taste: 'The spectacles worn by reason?'" in *Orientations* (Cambridge, MA: Harvard University Press, 1990), 44.

As Veit Erlamnn argues in *Reason and Resonance*, the mind-body split needs not be ignored, rejected, nor seen as nearly as split. We need "oscillation" between the two: to "know with our bodies and feel with our minds." Without this we risk a "loss of echo … the individual's dwindling capacity for self-reflection."[23]

23 Erlmann, Veit, *Reason and Resonance: A History of Modern Aurality* (New York: Zone Books, 2010), 315.

3

A BLACK HOLE THAT CONSUMES ALL DOUBT

4
Is Audio a Thing?

As the ways in which we interact with audio shift with ever changing technologies, the nature of what audio are for us has become confusing, elusive, and ultimately fractured into multiplicity. This is true for all mechanisms through which we relate to audio, from capture to production, to cataloging, to playback, to architecture, and spatial expectations for how to position our bodies in relation to it.

Our time is of multiplicity, of constellation, and of crashing through the ambivalent relativism of the last thirty years and crumbling attachments to highly capitalized unities. There is no competition between a possible experience and the original experience or any other, it's all an option, a temporary truth with specific possibilities and limitations that don't preclude the possible.

We can try them on, take them off, and remain unattached—use them to expand our lexicon of schemas.[24] Insofar as audio and listening are concerned, it is simply our job not to engage delusions that don't serve us, nor instead serve the last gasps of global media capital attempting to hand us directions to the ATM. It's too late now for any sole winner to arrive and control audio, and that's an opportunity for us to shape our relationship to it anew.

• • •

For the definition we will arrive at for audio, the question of an original moment's displacement, much discussed already, is—paradoxically—unimportant: a red herring that

24 A later essay, not published in this volume, addresses the term "schema," historically and for the refraction it can assist with in thinking about art.

distracts us from the full potential of audio and our relationship to it as a thing. But to arrive here, it was necessary to break our idea of audio in the ways attempted in the preceding pages. What we can accomplish in the words that follow is to portray several facets of where we can located the "thingness" of audio beyond the subjective realms discussed already, and in the recent history of audio representation as a spatial and social phenomenon in air.

We will intentionally avoid many facets and possible examples and misperceptions. For example, radio and wireless transmissions will not be addressed, punk and many other musics of the 1970s are not dealt with at all. The focus here is room-scale (or larger) audio representation technologies, their social architectonics, the social drive for their creation, and how they may have accidentally led us into a greater and richer awareness for how "audio" can be most productively understood. The intent is not historical completeness or depth, but to illuminate the history enough to clarify the broader concepts.

To give the reader a head start on where all this is going, I'll make a stab at a definition, but the etymology of "audio" is not particularly useful here. The Latin *audire* means "hear." This is not how audio function in our society. Our ears hear, we use our ears to engage audio, but we don't just hear audio, nor does audio hear us. Audio the thing is rooted in ideas and ways of thinking that have been embodied in technologies that provide a temporal displacement of a captured, transubstantiation of airborne sound that simultaneously offer the potential to desublimate this de-energized potential back into the air as a representation functioning as social architecture.

The inscription and reperformance of spoken language or music transcribed from performance and reperformed are both early instances of audio by this analysis. This is not to say that notation or written language in general is audio, but that these are both technologies used within a process by which audio are committed. *Audio is not a technology, but a process that employs technology.*

This process results in a place; audio is a thing, but that thing is both made of, and a maker of, place. The placeness produced (not the place of its making) can be considered a potential-original, but this original is not reproducible solely through an audio process, it requires a human being to "tune" place, with audio as one tool with which to shift the air utilizing a rare sensitivity and patience. Though difficult to achieve, it is not an impossibility to manifest a social architecture whose territory transcends the lexical and symbolic aspects of recorded sound and becomes a means for people to integrate these realms into something more, being together, or alone listening to ourselves listen. The realization of a location pregnant with this possibility is what is meant by place.[25]

The endless permutations of audio processes and their outcomes suggest that the word "audio" is best understood as plural. It is neither one thing or a category of thing, like "animal," nor is it a word whose contents can't be enumerated adequately, also like "animal." To generalize a definition for audio we need to look elsewhere, to the air and its qualities, through a historical investigation below.

• • •

My first memory of having agency within an audio process was the playback of a selection of my father's records, around 1985—I was five years old. The collection was housed vertically in an open shelving unit made of pine board and dowels. This meant that the records on the lower two shelves were easily handled by a small person, but the upper shelves remained entirely unknown. My dad is a hippie intellectual, long-term meditator, a psychologist; and the collection was more or less classics of late 1960s and '70s psych rock, jazz, blues, and a handful of non-European musics, Zen Meditation records (what do they do?), and a cluster of hammered dulcimer recordings from around where I was born in North Carolina.

Two of the records that caught my attention early on where The Beatles' *White Album* and Jimi Hendrix's *Are You Experienced*? The two records provoked very different types of experiences for me, but capture the beginnings, probably very similar for many kids, of a relationship with audio via representations of music, and the social space music (or its imaginary) can produce.

25 This desire to achieve a sense of place that goes beyond sound or music or audio or color or material or geometry is common to many artists and musicians, regardless of how they imagine the philosophical position of their work vis-à-vis in-itself qualities, a discussion we'll arrive at later.

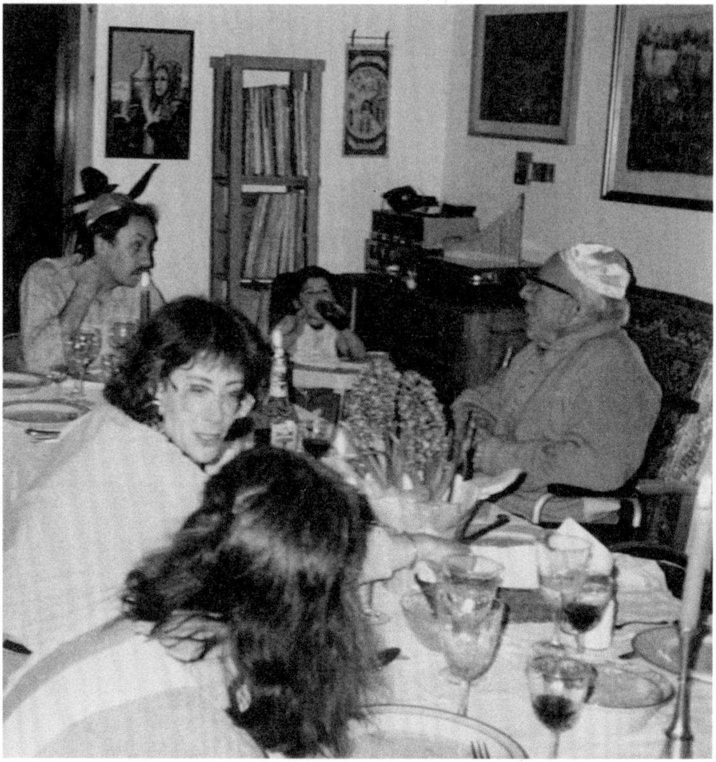

A living room with social architecture defined by family history and religious context, not audio.

The *White Album* I remember my father explaining to me. I think he felt it was critical that I knew it was an important record and had a special cultural value that until my later teens I wouldn't understand at all. Why was there nothing on the cover!?

Jimi Hendrix on the other hand quickly became all about transforming the room with sound. I would put on "Purple Haze" and jump around like I was setting a guitar on fire—a Hendrix story my father told me as I got into the album. Side two had "Fire." This record was the first instance of audio being a place for me, and this place had its own rules and was an invitation to certain types of radical behavior.

The room, when filled with the sounds of Hendrix's band, permitted actions that could be understood by everyone in the room as normal and acceptable conduct, but only while swimming in this particular, recalibrated air: room-saturating audio playback as social architecture. What had changed was an intersubjective relationship to the air and what was possible there, now induced by audio. Audio constructs momentary, affective physical behavior that is surreal, psychotic, or simply Other when the audio representation of sound is removed from sound.[26]

In the case of my Jimi Hendrix moments, the behavior felt only acceptable in the context of the audio's quantitative presence. Had I continued beyond the air produced by the audio, I would disrupt the normative behaviors I understood as required for that space if enacted without it. Dropping the needle initiated the representation, conjuring another social architecture. On a small scale, this temporary place is a site of liberation. At its best, this is what audio, in all of its inhuman, unreal artifact-ness can do for us.

• • •

Capturing the total intrigue of historic relations between audio and conduct is beyond the intent of this text. For our purposes, we will sketch the evolution of building-scale and what can be called "Land Art-scale" sound systems from the late 1960s to present with a focus on the US and Europe. This is where the majority of the technology was developed before Japan (and now China) took over the manufacturing of personal audio, introducing the Walkman in the early 1980s.[27] We also will intentionally not focus on the

26 In the artist Matthew Barney's *Cremaster 3*, two hardcore punk bands perform in the Guggenheim Museum as part of "The Order" segment. They are on stages made of salt and in front of them is a mosh pit surrounded by security guards. The bands are at times entirely unamplified: power chords with no power, just the small sound of a pick and a solid piece of wood, screaming vocals nearly inaudible. But the social architecture, a ritualistic mosh-pit situation remains unchanged and the effect is an extraordinarily rendering of this phenomena. A distant, but related example: if we could manifest in space the imaginary conversation partners of a schizophrenic, would the individual still be schizophrenic? Can the words produced by psychosis be considered audio in a culture where a state of mind is othered by a notion of disease, as if a distinct entity is using the body as a puppet?

27 Japan was a major manufacture of home hi-fi components throughout the 1970s, but the design innovations were primarily American and European.

music or sound recordings themselves, as much has been written about all of the artists and movements mentioned. What have received much less attention are the situational relations between audio technology and social space: large-scale, communal topographies of air. This will take us toward more possible clarities in thinking about how to locate audio.

• • •

La Monte Young's *Dream House* is located in lower Manhattan and has been, with some interruption, an ongoing sound installation for more than twenty years. The idea for the space was conceived in the late 1960s. The upstairs rooms facing Church Street contain a magenta light installation by Young's partner, Marianne Zazeela, plush carpeting, and specially created loudspeakers whose main drivers are nearly four feet in diameter. An analog synthesizer tuned to thirty-two different frequencies is creating the composition invisibly. The result is a listening situation, in collaboration with the architecture, design, and lighting, which seems to be an unchanging audio environment, but only if you are still. As the space invites you to move, with its differing zones of light, the frequency lattice reconfigures itself as you enter different nodes of the various wavelengths being projected and your body absorbs a new set of frequencies, the ear presents a new, coloristic distortion.

The room and the visitor's body are used as an elaborate four-dimensional filter. This work is perhaps the most widely known and lauded example of "sound installation" post-1960, and aside from Max Neuhaus' *Times Square* (which began earlier and has run longer, under the auspices of Dia), the most "permanent."

One would think that a work like this would be deemed beyond syndication, but several museums have at great expense attempted to recreate the *Dream House* as part of survey exhibitions relating to "sound art." We'll consider two of them in addition to the Church Street space: Centre Pompidou in Paris, and Zentrum für Kunst und Medientechnologie (ZKM) in Karlsruhe, Germany. In both cases, the installations were prepared by the artist and a top technician, who has worked with him for several decades. Nevertheless, if we consider the imperative of the work in

New York, the attempts to relocate the *Dream House* were failures. These attempts at reproduction present a unique paradox to lance.

Young's *Dream House* is a place that involves an audio process, but is not primarily audio. Audio is but a component, and one that is contingent on the subtle play of architectural acoustics and unfettered control over sound levels. In New York, Young owns the building and lives there on a separate floor. Loudness is not mediated by consideration for other tenants, for the autonomy of other artist's work, or for a sense of decorum germane to friendly and state-funded cultural institutions. The necessary preconditions for the work's installation rarely exist in a museum setting.

Were he to situate his practice and his notion of an original, or "the work," differently, Young could have chosen to recalibrate the audio frequencies to accommodate the new acoustic and social conditions; he had the option not to commit an act of audio against his own piece, which in New York is actually an instance of music much more than audio. But he believes that there is an original in the form of a place, and that this can be disentangled from what he accomplished uniquely within a building he owns and whose context he has cultivated over many years.

For Young, the frequencies chosen for inclusion in the *Dream House* have their own, intrinsic property that can function aside from their contingent effect as air. This articulates his refusal of this contingency and replaces it with a universal notion whose philosophical basis lies in the time, place, and affect specifically within several strains of traditional music in India, realigned in collaboration with Young in the 1960s by the mathematician Catherine Christer Hennix. On the one hand, the world claims a specificity few other American or European-continuum musics have claimed since the Middle Ages, yet simultaneously rejects an empirical difference obvious to anyone relating an experience of Church Street to ZKM, for example. Young's willingness to relocate the work betrays the primary insight of his work: music's potential to transform consciousness by tuning the body via listening.

So when does a place produced in part by audio become or not become this "is" rather than the "ism"[28] of a

28 This is a rearrangement of a phrase Anthony Braxton used often while teaching at Wesleyan University in the 2000s: "Don't let the -ISM get bigger than the IS."

failed attempt at reproduction, yielding an unintended representation aspiring to be a lost original? If our perceptual apparatus works within a bandwidth of sensitivity and history, unique for each of us, how is it that a large group will "know" something is not happening that could be? There is a sense of loss that comes with this—something has not taken place that should have. We might expect this perception to be quite individual and the product of taste alone, but it is often felt clearly by groups as well. We have all attended a performance where, aside from whether we enjoy or don't enjoy it, it is clear that a perceptual measure of it being present was somehow not achieved.

Perhaps it is through a kind of perceptual mirroring of the air that we find the inner and outer merging; we have a sense of a close encountering of present-placeness within a duration. We know what this feels like, but how is it that this merging into a shared sense of the moment is a kind of precondition for having peak affective experiences? This is not only true of music or performance or an artwork, but also of friendships, love, and sexual encounters.

The evolution of audio technologies has been an attempt to manufacture this sense of being together here and now in a reliable way. In music this is produced via a transpersonal, delicately human capacity often thought of as a very rare occurrence within the neurodiversity of the species. Attempting to machine the air into the byproduct of this interhuman form of communication and alignment is particularly true of loudspeakers: aside from lightning they are the primary interface between electricity, air, and our bodies.

• • •

Audio is Air is Social Architecture

"Anything less of what we are capable of perceiving is an insult."[29]

In the earlier story of childhood desublimation, in which ecstasy was facilitated by Hendrix and a parental sound system, the social architectural rupture was not inhibited by any lack

29 The Funktion 1 Sound System Explained, streaming video, AIAIAI, 2012, http://www.aiaiai.dk/blog/the-funktion-1-soundsystem-explained.

of sensuousness in the representation. At that age, few children, this one included, had heard other home audio systems; nobody in the family or community was particularly audiophilic, and the family weren't regular attendees to venues for music that would have had paid any close attention to, or financially invested in, the details of their sound systems.

However, it is the sensuality of the air, even in the form represented in that living room that seized a child from the mundane. This type of abduction into a heightened state of being is what La Monte Young produces at *Dream House* and what artists such as Maryanne Amacher and Iannis Xenakis (in his polytopes, e.g. *Persepolis*) each developed in their own way in the 1970s within the area of experimental music.

Beyond the living room there was extensive exploration of what audio could be when projected at large scale. The historic Groupe de Recherches Musicales (GRM) in Paris formalized their orchestra of loudspeakers called The Acousmonium in 1974.[30] This sound system was used in conventional concert halls with the speakers (many different types) placed on stage as if to simulate the spatial distribution and diversity of radiant patterns of instruments in an orchestra. It coincided with a performance practice to "diffuse," or to make more spatial, music composed on magnetic tape. At the beginning of this practice, the tapes used were monophonic—no spatial information was encoded (as is the case with stereo or multichannel audio formats). Spatialization was achieved via the spectral information's relationship to the loudspeaker chosen by the performer, mixing techniques, as well as the architecture. This audio-related performance practice has continued via academic programs in electronic music throughout Europe and Canada, with some presence in the United States, Argentina, Mexico, Brazil, and Japan.[31]

30 This system's intention was to replace and extend performance positions on a stage for the live "diffusion" (mixing) presentation of prerecorded magnetic tape.

31 This practice in North America has been most prevalent in Montreal where the shared language (French) afforded extensive back and forth between Canadian and French composers. In Mexico, Argentina, and Brazil there are a handful of universities with small departments that engage in this practice as well, though the continuance has been European. France, Belgium, UK, Germany, Austria, Switzerland, and Spain each have educational and cultural institutions that still support this practice of live diffusion.

In the same year The Grateful Dead went on tour with their "Wall of Sound." Preceding these developments in the United States and Europe was Jamaican sound system culture, whose participants had already for a decade been making homemade sound systems

see page 8

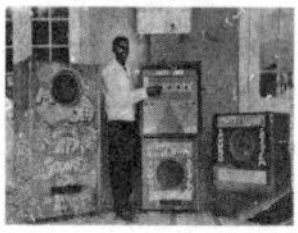

that could project audio at Land Art scale. In Jamaica, the production of recordings, performance with these systems, and the systems themselves were often owned by the same person. A "sound system" in this context doesn't only refer to the audio equipment, but to the people involved as performers and engineers as well. The human-technology amalgam as "sound system." But the technology was central: often extremely powerful at 20,000+ watts and of very high quality, "sound systems" performed outdoors, the dance hall was the city.[32]

Contrary to how we listen to dub today, largely as personal audio played in our homes or cars, the early records played in these outdoor spaces were produced solely for this practice and responded to weather-condition diffusion. Even the discs themselves were made out of an extremely soft material that would not decode after many plays—although dub was and is a "studio music" that produces recordings, it's origins are not coherent with an American or European conception of record distribution. So, when we listen to Mad Professor or King Tubby or Augustus Pablo now, we should think of this as if we are experiencing a Smithson via fax.

What made all of these, and many other new ways of thinking and experiences, possible were major imaginative leaps and subsequent technical developments in loudspeaker system design. Beginning in the late 1960s, loudspeaker, amplification, and signal processing technologies coalesced around a new vision of detail, quality, and scale, which made possible large-group social contexts within which experiences of the new air these systems became possible. These situations isometrically emerged in nearly every area of musical and stage production, and traveled quickly around the world via excitement over what they could accomplish socially.

The first large rock music festival in the way we think of them today happened only in 1967: it was the Monterey Pop Festival. The sound system for Monterey was designed

32 Veal, Michael, *Dub* (Middletown, CT: Wesleyan University Press, 2007).

by Abe Jacob, who had been a Bay Area engineer for rock shows (Jimi Hendrix and Peter, Paul, and Mary, to name just two). The system was minuscule by today's standards, but was an enormous leap from what had been the norm: bands performing with only vocals (and perhaps some drums) through a tiny on-stage PA. Individual instrument amplifiers would provide the rest of the direct sound. For example, it was only three years before Monterey that the Beatles came to the US for the first time, playing with thirty-watt VOX guitar amplifiers and a one hundred-watt public address for vocals. Today, most small cars come standard-equipped with nearly that much power.

see page 26

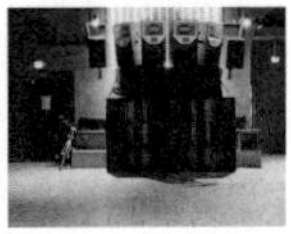

In the image below of the Monterey Pop stage, we can see that there are two stacks of speakers on stage and two stacks rigged on the top of the stage roof. At this time there were no digital delays, so we can assume that the time delay between these was audible, although it did ensure that the throw of the sound would reach the entire crowd. There were no fill-in speakers or surrounds, as we'd see today.

With Monterey we see the American beginnings of a reciprocity between a social desire for collectivizing audio experience (though it was thought of as live music and not audio) and audio technology that would need to quickly evolve to meet this desire. In this initial shift away from each musician directly controlling the sound reaching the audience (as the Beatles and everyone else did in this period), the model of audio as a representation and not an "original," even in a live music setting, comes into focus.

see page 12

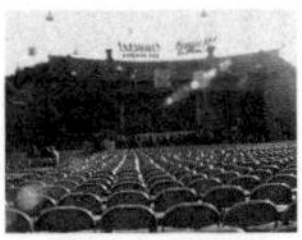

The Grateful Dead's "Wall of Sound" system was a near-term revolutionary return to the previous era in that it did away with the need to have a mixing engineer in the audience creating a duality of perception between the band and the audio representation—common practice then and now. The band on stage listens, reacts, and shapes a representation of their performance for themselves in collaboration with an audio engineer (often in addition to the house engineer) via on-stage speakers that face the musicians. The audience hears a mix composed for them by an engineer who shares their listening perspective.

If the intention of a concert at this time was to create a collectivizing, transpersonal experience that bridged the experiential gap between performers and audience, everyone involved in that process should be engaging with the shape of one representation, not two. The Grateful Dead's approach was a Utopian model for audio that reflected the band's cultural position, but ultimately, the transport and setup costs of such a system proved financially unfeasible.

The designer of the Monterey Pop system, Abe Jacob, is also credited for inventing the field of sound design for Broadway shows (he did the original runs of *Hair*, *Jesus Christ Superstar*, *Cats*, *A Chorus Line*, *Chicago*, among others). He moved to New York City in order to manage Hendrix's recording study Electric Ladyland, on the basis of his reputation coming out of Monterey. He happened to be in a theater while there were some technical problems, which paved the way for his perspective to spread.[33]

Prior to his presence in the field, theater sound was simply handled by the stage manager, who would not have had the expertise to calibrate the experience of the room in the way a specialist would. At that time an audio engineer nor sound designer was a default component of a theater's operational staff.

The production of theater now, with performers often wirelessly mic'd and mixed from the audience, along with lighting and video automation, is recent and a radical shift from the idea which has existed for over a millennium of the theater not as a media technology, but as a stage with supportive acoustics for people. The theater today is an assumed mix of media, technology, and human capacity; a total media instrument—a trajectory put into motion by the chain reaction caused by Jacob's transport of '60s large-scale rock-context audio innovations.

At this time, every large-scale PA was a custom-built enterprise and still relied on theories of sound diffusion physics that were extensions of the movie sound speaker arrangement (horizontal or small vertical stacks on stage). Further, the outdoor venues used for rock and pop concerts had almost exclusively been designed for orchestras performing without amplification. The architecture was designed to project

33 Interview with Abe Jacob, *Live Design*, September 1, 2002.

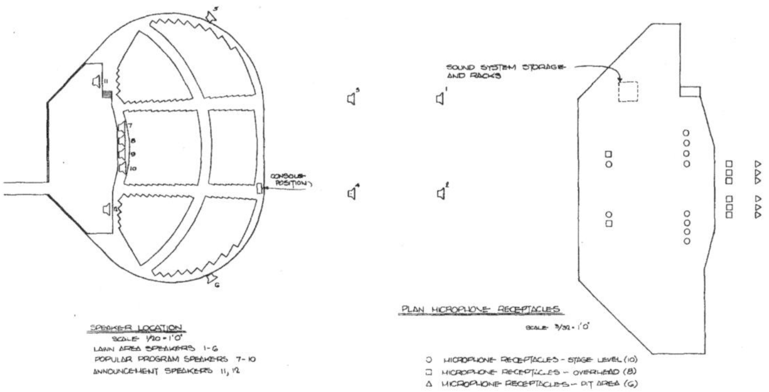

Loudspeaker placement diagram for Mississippi River Festival, 1969

the acoustic mass of an orchestra outward to an audience of two thousand or more, attempting to simulate an indoor concert hall.

The audio engineering community was struggling to find a way to overcome the disconnects between recent and established architecture built to reinforce acoustic sound via reverberation and amplified sound, which would destroy the clarity, especially at the high levels desired. In a 1969 paper presented at the Audio Engineering Society's annual convention these issues were described:

> **No sooner had we resolved the problem of providing a suitable environment for natural orchestral sound in these buildings ... then, in typical American fashion ... the program use broadened and the very spaces that were planned only five or six years ago, solely for orchestral use ... became stages for a wide variety of popular music attractions, including rock and roll.**
>
> **Suddenly, we were faced with a series of basic conflicts in terms of providing an excellent natural orchestral sound in the pavilion, high-quality symphony reinforcement on the lawn, and intelligible pop and rock sound, at high levels, in both the pavilion and the lawn areas.**

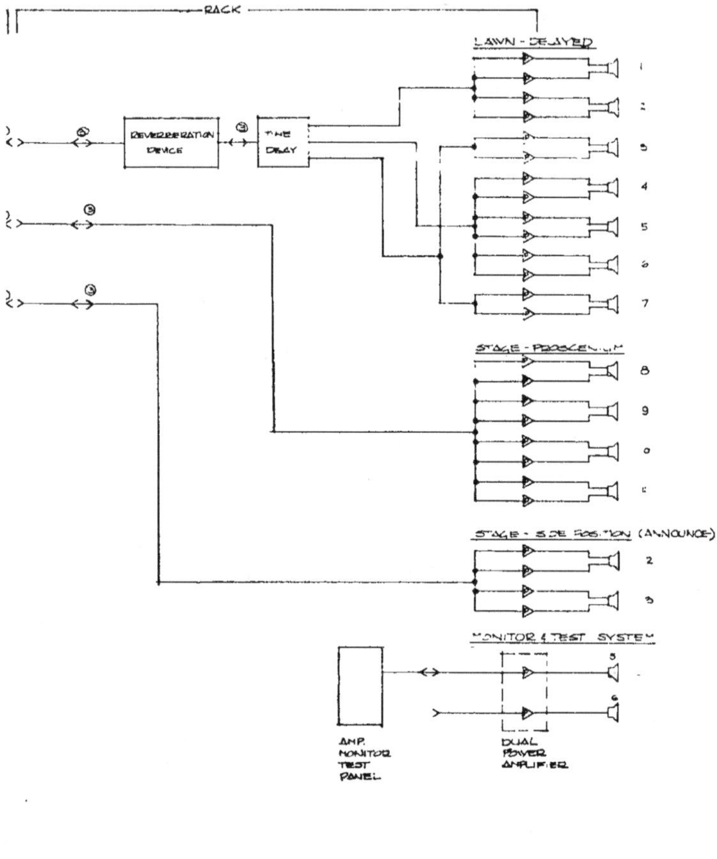

Diagram describing an early time-delay system used at the Mississippi River Festival for aligning the arrival of acoustically produced stage sound and that of the audio system to audience members in different locations.

> **The very physical acoustics techniques that were employed to create sustained sound pressure levels of the reverberant field in the early portion of the decay curve, came back to haunt us as unintelligible mush when the rock groups' guitar amplifiers performed in the orchestra shells.[34]**

The paper presents a design for the Mississippi River Festival in St. Louis. As you will see in the images at left and on the previous page, there are already major innovations to what existed at Monterey. Speakers are placed at several distances with a time-delay system employed so that the bleed between layers of amplification would be reasonably time-aligned.

Throughout the 1970s, significant technical and production innovation forced a paradigm shift to take place. The major manufacturers like Altec, ElectroVoice, and MGM who developed their business models, research approaches, and designs during the movie sound era (and for sound reinforcement needs associated with acoustic country and jazz) were unable to anticipate the cultural shift that was underfoot, and how it related to audio representation in new social spaces. Most were slowly outrun by younger engineers who were a part of the new vanguard of cultural life in their generation and invested in what these systems needed to be capable of achieving.[35]

• • •

Less than a decade after the explosion of rock music as a style—and as an experience of large-scale audio—disco emerged. From the beginning, disco was, as the arts writer Tan Lin described it, "an operating system."[36]

Disco was as much a place as a style of music, and these places mushroomed throughout the US, Europe, and South America. The popular movement unfolded from loft parties in New York City in parallel development to the reappropriation of

34 Jaffe, Christopher, "Sound Reinforcement in the Music Pavilion," in *Audio Engineering Society Convention* 37 (1969), accessed March 21, 2013, http://www.aes. org/e-lib/browse.cfm?elib=1316.

35 About Meyer Sound, accessed March 21, 2013, https://meyersound.com/about/

36 Lin, Tan, "Disco as Operating System, Part One," in *Criticism* 50, 1 (2008), 83–100.

industrial buildings by visual artists and musicians in lower Manhattan, usually referred to as "the loft scene." These early spaces are the whistle that begins an ongoing cultural moment, where extremely detailed audio representation at building-scale, becomes the defining necessity for certain types of peak, affective, communizing experiences.

The experience of audio produced by the sound systems of upper tier disco clubs was unlike anything a human being had encountered before. These places were the convergent sites of social desire, technology, engineers, artists, and perceptual modifications afforded to participants primarily through pharmaceutical stimulants.[37] We should not underestimate the ripple-effects of such spaces for transforming the senses and the unpredictable cultural shifts theses sensory developments can propel.

Before continuing to explore the air of disco, there are three sites where other spaces of remarkable audio could have been experienced prior: the "movie palaces" of the monopoly-era Hollywood film industry; recording and mastering studios; and in Jamaica (or among the Jamaican diaspora) engaged in sound system culture.

A now-distant fact of audio history is that from the 1920s until the early 1950s, the highest fidelity recording medium was film.[38] It had the least noise and greatest clarity. In cinemas of this era, one would have experienced audio played back via this medium, far superior to radio or any format within the evolution of phonography. However, the acoustics in these spaces were not tuned for recorded sound, but were largely designed from a vision of exotic architectural simulation—built around Egyptian, Chinese, or Aztec fantasy, or in simulation of European grandeur. An exemplar is the Aztec Theater in San Antonio, Texas, whose original proscenium decor was imagery of Montezuma meeting Cortez.[39]

37 Lyttle, Thomas and Michael Montagne, "Drugs, Music, and Ideology: A Social Pharmacological Interpretation of the Acid House Movement," *International Journal of the Addictions* 10 (Oct. 27, 1992), 1159–77.

38 Mosely, John, "Motion Picture Sound in Record Industry Perspective," in *Audio Engineering Society Convention* 66 (1980), accessed March 26, 2013

39 Dumbarton Oaks Ephemera Collection, AR.EP. PC.0670. https://www.doaks.org/research/library-archives/dumbarton-oaks-archives/collections/ephemera/procenium-arch-and-organ-grid-aztec-theatre-san-antonio-texas

For the duration of this era, the loudspeaker systems would be located behind the perforated screen, and were not intended necessarily to fill the space with sound, as would later become the norm. After the acoustic properties of disco and immersive sound were appropriated as spectacle by the culture industry as a whole, cinemas transformed into the acoustically dead, high-decibel audio rooms they are today.

see page 6

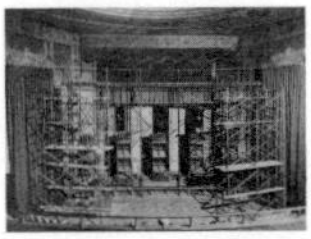

Within the recording industry, since the 1950s, there were studios whose sound systems could provide extraordinarily detailed and powerful audio representations, but these were cloistered locations, exclusive to professionals and their invitees, and only available to small numbers of people at once. This remains true to the present, where few people, even music professionals, ever enter a precisely tuned mastering studio where the build-out costs can be in the millions. In Europe there were also electronic music studios popping up, particularly in France and Germany. These spaces, and the composers who were afforded the luxury of treating them as near-private laboratories, were heavily invested in by the state. This investment came in part from a desire to reidentify a national music and identity after World War II, as well as to participate in the scientific and technological optimism of the time.

In France, Pierre Boulez was able to initiate IRCAM, which is to this day probably the most well-funded center for electronic music research. The blaze of postwar scientific optimism afforded Boulez the change to institutionalize an incredibly esoteric and cloistered area of artistic practice in concert with a deliberate ignorance and public suppression of its military implications and the funding contingencies of complicity.[40] (This quiet symbiosis continues into the present with the Defense Advanced Research Projects Agency [DARPA] funding university-based media art institutions such as Stanford's Center for Computer Research in Music and Acoustics as well as MIT's Media Lab.)[41]

40 Born, Georgina, *Rationalizing Culture: IRCAM, Boulez, and the Institutionalization of the Musical Avant-Garde* (Berkeley: University of California Press, 1995), 159.

41 Wright, Matthew, et al., "CCRMA Studio Report," accessed May 5, 2013, https://ccrma.stanford.edu/~matt/studioreport/CCRMA- studio-report.pdf. Gordon Wetzstein et al., "Compressive Light Field Displays," in *Computer Graphics and Applications, IEEE* 32, 5 (2012), 6–11.

Meanwhile, back in New York and not far from Bell Labs and the first phases of Experiments in Art and Technology (E.A.T.), disco was unfolding.

• • •

Back to Disco

In New York, the combination of harsh postindustrial acoustics, a desired social architecture, and economic necessity on the part of the clubs to meet these desires or fail pushed technology to develop quickly. The major innovations of this era were numerous.

Richard Long & Associates (RLA) designed many of the top clubs during the disco era, notably New York's Studio 54 and Paradise Garage, Ministry of Sound in London, along many others around the world. RLA is broadly credited for making discos the immersive audiotopias that they were.[42]

The reason RLA became a worldwide name was that each of their systems was thoughtfully and painstakingly tailored to the individual needs of the architecture and its distinct use. The process was somewhere between that of a high-end tailor crafting a bespoke suit for a VIP customer and a specially commissioned artist painting a portrait of an esteemed patron.[43]

This "tailoring" created rooms that were hypersensitively aural in ways the general public had never experienced, and the most advanced DJs of the era were exploring the potential of this

see page 6

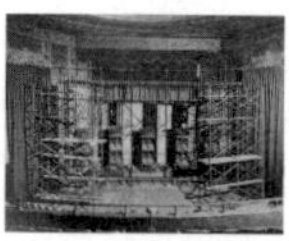

sensitivity with sophistication. A social architecture emerged around liberated sexuality and identity within the semi-autonomous, near-imaginary air that the disco sound environment enabled. Long was not simply an engineering bystander, but an active participant in the imaginary. As one playful indication of this, he named his inventions in ways that linked them to the social milieus they engaged: Double 12" Dildo mid-woofer (there was

42 Kun, Josh, *Audiotopia: Music, Race, and America* (Berkeley: University of California Press, 2005).

43 Richard Long & Associates: "The Men Who Made Disco Go Boom," accessed January 24, 2013, http://societeperrier.com/new-york/articles/richard-long-associates-the-men-who-made-disco-go-boom/.

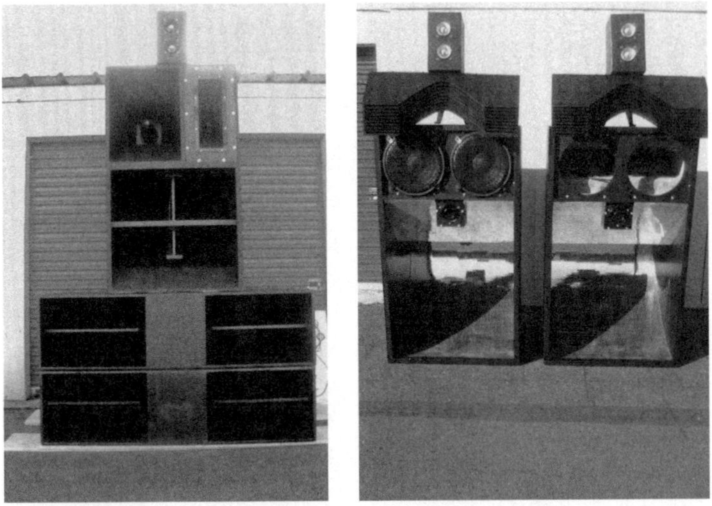

Richard Long-designed loudspeakers from Zanzibar, in Newark, New Jersey, one of the top clubs of its time.

a giant plug that could switch the phase which looked like a dildo) and Puissance (French for "power"), for example. It is impossible to imagine Meyer Sound in the US or D&B in Germany understanding their role as aligned with one or another social movement in this way. Instead these companies furthered the Fordist modeling of technology which yields concepts of media transparency and universal utility rather than hand-spun and locatively invested.

Paradise

RLA designed the Paradise Garage system in use at the time. It is illuminating to take a close look at this system, as it was arguably the apex of disco sound in North America (though many would argue that by its opening in 1977, disco had been largely co-opted by the record industry and had a transformed clientele and weakened cultural position). Regardless of disco's cultural status, the Paradise Garage as audiotopia benefited from almost a decade of rapid technological advance.

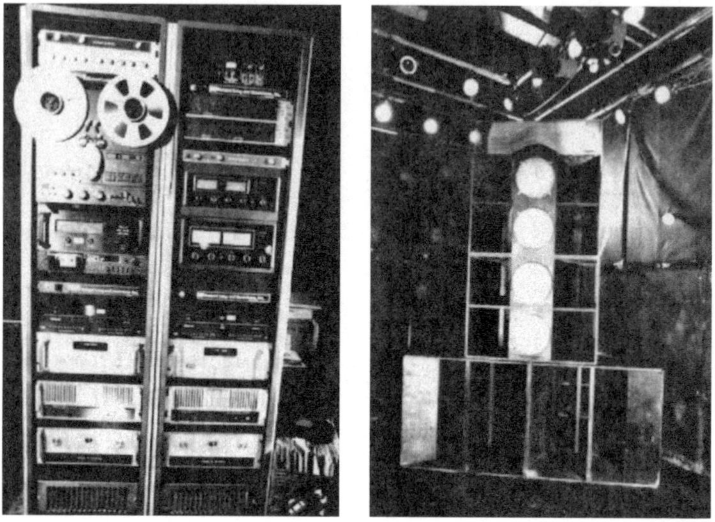

Left: Equipment Rack at Paradise Garage
Right: Levan Horn (bottom) and Ultima (top)

Previous to opening the new venue it had been another dance club. Enormous, it was a 20,000-square-foot space with a single dance floor taking up the majority of the space. In its earlier iteration, the business had been a financial disaster and closed. According to Long, the failure of the business was due to a sound system that was extremely undersized for the room[44] and consequently was out of step with the expectations of disco patrons of the time.

An ex-lover of the Garage's owner (Mel Charen) put up the money to retain RLA and ensure the sound system was ideal.[45] RLA agreed to install the equipment over a period of time as funding could be delivered. The firm phased out the old system in pieces, developing custom devices to suit the room. The resident DJ would be Larry Levan, who was among the most widely

44 Fierstein, Alan and Richard Long, "State-of-the-Art Discotheque Sound Systems-System Design and Acoustical Measurement," Audio Engineering Society Convention 67, 1980, 1.

45 Pareles, Jon, "Paradise Garage, a Gay Club That Forever Changed Night Life," *New York Times* (June 18, 2000).

acclaimed of the time. The sound system would be developed not only to address the acoustics, but also Levan's performance technique, cultivated over the previous decade.

The new owners subdivided the warehouse and the main dance floor was reduced to 5,000 square feet. For this space, Long designed and fabricated a unique bass horn (called the Levan Horn), as well as a new enclosure to replace the four existing sub-bass speakers. Each was 1000 watts and had a twenty-eight-square-foot "mouth" (the open surface area that projects sound outward from the drivers). According to RLA's testing, one of these overpowered all four of the speakers previously used to activate a space four times the volume.

A unique full-range speaker was also designed (the Ultima) and only the original overhead tweeter arrays were retained, though used differently within the overall signal flow and control system. RLA also designed a unique crossover system in collaboration with the acoustician Alan Fierstein whose "most unique feature is that the two extreme ranges of 20 to 100 Hz and 7K to 20K Hz are controllable in volume by the disc jockey with up to 16 dB of gain built into the circuit." (For reference, every increase of 6 dB is a doubling of perceived loudness.)[46]

Against criticism from his professional peers, Long defends the choice to give the DJ direct control over such a powerful sound system:

> **In order to explain our concept of a disco system, let us give this analogy: In a discotheque the sound system can be considered to be the orchestra while the DJ is the conductor. The conductor's job is to stimulate and entertain the audience; the DJ must entertain the dancers. The DJ is not reproducing the works of Bach or Brahms as performed in a symphony hall, but is instead playing music which was created in a multi-track studio under artificial conditions and mixed by an engineer also attempting to create the most exciting sound possible.**[47]

RLA developed many other special components for this space, including a suspension system for the turntable tuned to 2HZ with rubber

46 Fierstein, Alan and Richard Long, "State-of-the-Art Discotheque Sound Systems," 1980, 3.

47 Ibid., 2.

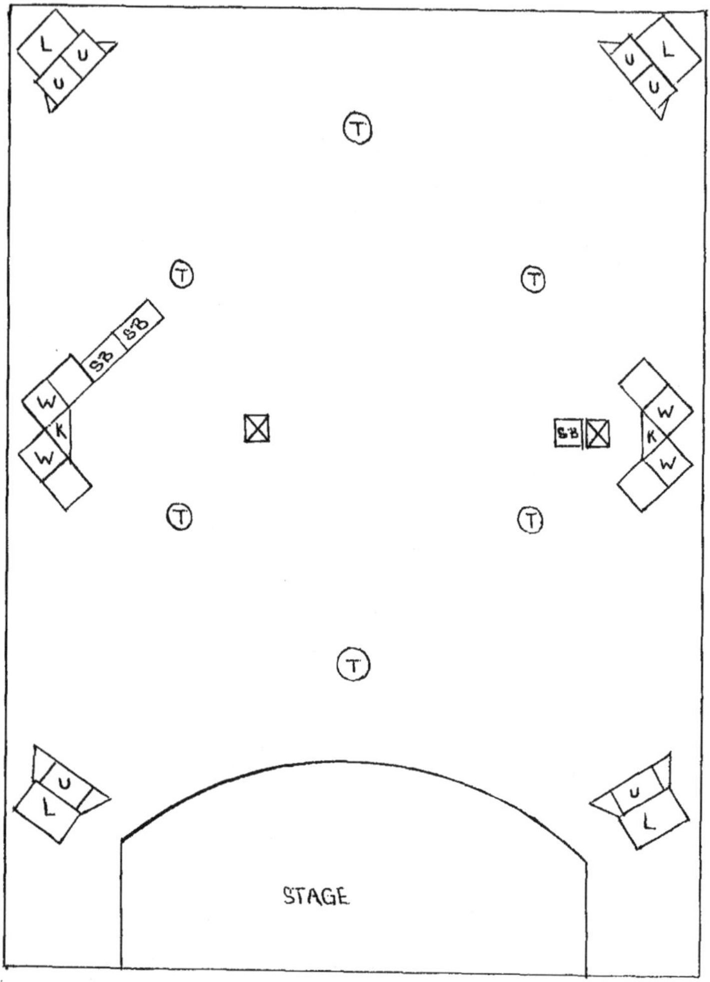

Figure 4- Speaker layout at the Paradise Garage.
L= Levan Sub-Bass Horn
W= Waldorf Bass Speaker
U= Ultima Full-Range Speaker
K= Pseudo-Klipsch Horn
SB= Small Sub-Bass Horn
T= Tweeter Array (hung from ceiling)

Speaker layout at the Paradise Garage

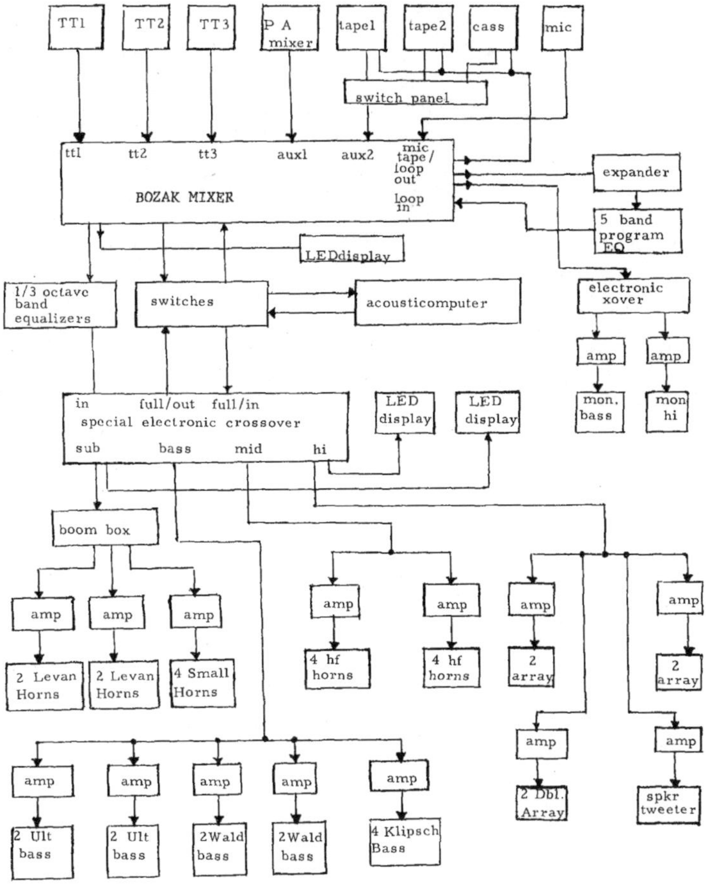

Paradise Garage sound system block diagram

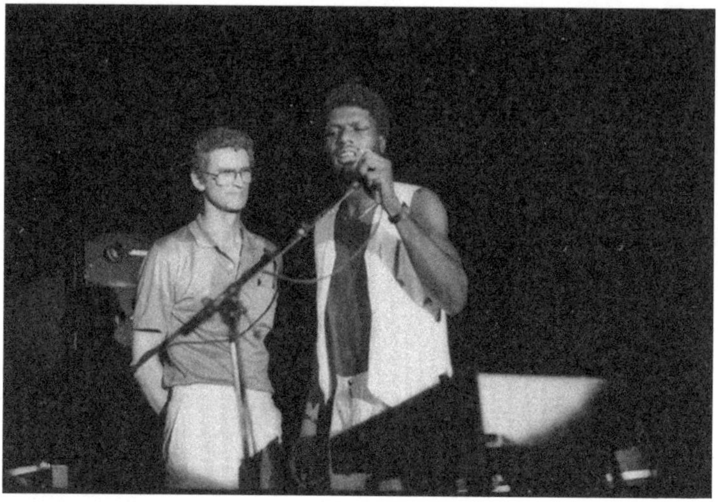

Richard Long and Larry Levan

bands. He viewed their role as acoustic designers support system for the artistic needs of the DJ (here we can hear echoes of GRMS's approach to the creation of tape music tools and the Acousmonium):

> **Since the DJ is responsible for creating an exciting sound, we try to make sure he has enough tools at his disposal. Such special effect devices are:**

1. **Our special Electronic Crossover (discussed earlier);**
2. **The DBX Boom Box, which provides a blend of 25–50 Hz bass synthesized from 50–100 Hz information present on the recording**
3. **Dynamic Range Expanders, used to undo the compression found in most recordings**
4. **The Deltalab Acousticomputer and similar devices, used to alter or add to the sound of the recording**
5. **The Audionics Space & Image Composer, a 4-channel synthesizer**
6. **New devices currently under development, such as the Acoustilog Image Enhancer, which expands the stereo effect.**

All of these devices, except the last two, are currently installed at the Garage.[48]

Following is a listener's description of how Nicky Siano, another of the foremost DJs of the time, was thinking about the social architecture, silence, and reactions to different frequency bands. Siano would have been using similar controls to those described above:

> **Nicky Siano took his sub bass beyond. He wanted this heavy sound.... his Richard Long tweeters, he had Alex Rosner stuff also, but he wanted the crossover points to be so much that when you took out the center, you could hear no program [i.e. song], you could hear tss tss tss [hi hat], and bummmm bumm [bass] and every once in a while, especially in the black clubs, with the dancing, everybody would be so together with it, they'd be sometimes singing and Nicky would bring it up and then all of a sudden BOOM! It was flawless, and the lighting was with it perfectly. When the room went to clack [i.e. the music] it would be black [the lights]. And yet there'd be a couple of hundred people in there dancing. And you couldn't see your hand in front of your face. And it'd be so intense. And then all of a sudden something would come on. Then lights would come on like normal, and everything was dddddd [climactic brass stabs]. Nothing automated. It was perfection.**[49]

Audiotopia

The social architecture of disco was an outgrowth or even the more real, realization of the '60s optimism via social dancing in a hyper-affective space. While arguing that this would not have been possible without parallel developments in audio representation, the two (just as rock festivals had several years earlier, and urbanization drove the realization of Jamaican sound systems) needed each other. Building-scale sound meant that a small town's population could

48 Ibid., 5.

49 Broughton, Frank, "Bob Casey, NYC Disco Pioneer," http://daily.redbullmusicacademy.com/2016/06/bob-casey-interview.

fit themselves onto the floor of a warehouse and quite literally be swimming together in sound waves. This meant all bodies in contact, moving together, transcending the common kinetic limitations of individuality present during the 1950s and early '60s America of the dancer's youth.

Discos have been historicized as a dominantly gay scene, and while it is clear that in post-Stonewall New York City discos became homes of gay liberation, it was also a home for a social context that could absorb any willing body into its territory. This communality is also what defined early rave culture, which is the next occurrence of large-scale social dancing after hard rock and pop-spectacle overran disco in the Reagan years.[50]

Air Everywhere

Land Art-scale audio quickly expanded into every cultural layer of society. In 1975, Accentech, an acoustical consultant, was hired to make a permanent installation at Tanglewood.[51] What is now known as the Koussevitzky Music Shed is about as culturally distant from The Paradise Garage as you could get in 1975, but its important to link these developments as the technology behind new social architectures.

At this time, with amplified orchestras only beginning to exist, it wasn't clear that a public would pay to sit on a lawn, beyond the enclosure for the orchestra, and listen to a representation of the music being performed out of earshot. This new type of "being there" is an entirely different aurality and sense of human connection between musician and audience than had existed in orchestral contexts before this.

What is experienced is not an "original," even with the physical proximity to the performers, and to go a step further, without a direct acoustic link between performer and listeners' bodies, we can ask ourselves: is this experience even "music" in the way it was understood anymore?

50 Disco and the Queering of the Dance Floor, 2013. Accessed January 24, http://www.academia.edu/1066547/_Disco_and_the_Queering_of_the_Dance_Floor_.pdf.

51 Anne Guthrie, Acoustical Consultant with ARUP, personal communication with author.

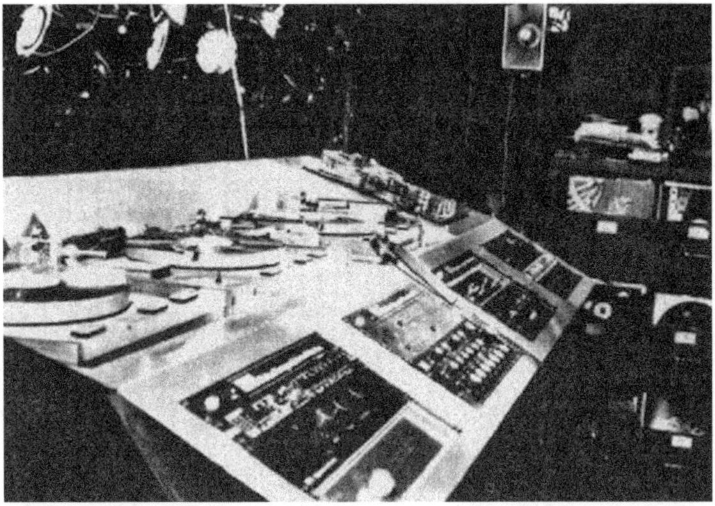

DJ controls at Paradise Garage

Is a powerful loudspeaker system significantly more distancing to a direct human to human musical encounter than a cathedral-scale organ's pipes? What if we were listening to the cathedral amplified to an adjacent lawn?

These distinctions are not important for our project, but the obvious complexities of the many possible scenarios we can imagine and how little resolution beyond rejection/acceptance we have in our culture. The readiness to accept these types of representations as music paved the way for the confusion we now have to identify what isn't audio, rather than what is, and our reliance on hyperbolic descriptions of musical style or genre as represented by a canon of recordings, rather than anything more absolute.

The Sound of Composition's Aura Being Popped

The period of time from the late 1960s to 1980 we've been thinking about in terms of loudspeaker systems is simultaneous to shifts in understandings of how notions of authorship morph under the influence of the new technologies. While overdubbing was already

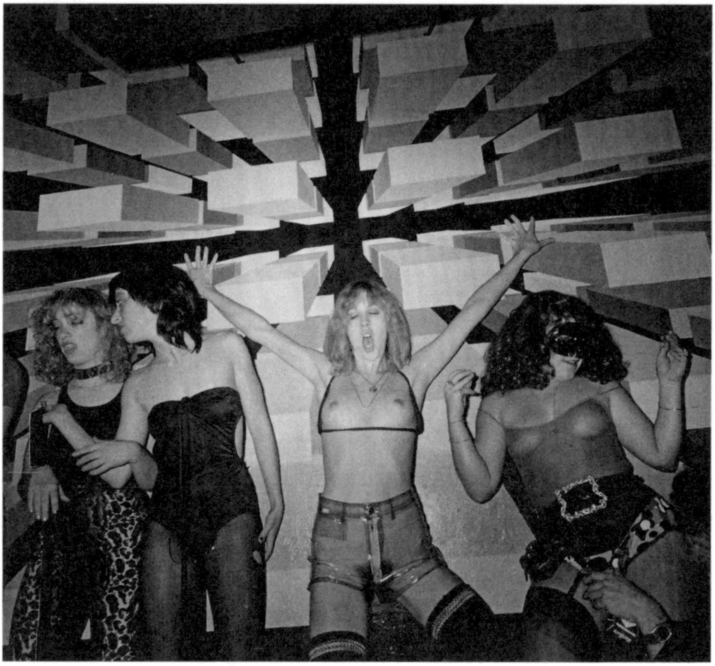

Dress code at Infinity in 1978

a practice in the early 1930s,[52] it is the transition to magnetic tape-based audio that afforded what we think of now "the [recording] studio as compositional tool."[53]

As tape gave way to digital in the 1980s, algorithmic composition tools—software-based composition systems that were not simulating the recording studio, but were explicitly "compositional tools"—could permute music endlessly and became dominant in the industry. While algorithmic models were already being used on paper by composers in the 1960s, entirely

52 An Enrico Caruso recording from c. 1910 was overdubbed with an orchestral accompaniment and released in 1931 by Victor.

53 Brian Eno, "PRO SESSION–The Studio As Compositional Tool," *Downbeat Magazine* (July 1983). Of course this quote from Brian Eno was old news already to practitioners, but Eno was in the unique position to popularize an idea that had been enacted in the cloisters of academic "tape music" produced in Europe and US beginning in the late 1950s.

computer-based, real-time sound synthesis originates with Max Matthews at Bell Labs. This practice remained relatively cloistered throughout the 1980s and exploded in the '90s as personal computing grew more powerful and inexpensive, and the computer industry embraced sound cards as a necessity.

While electricity made possible the first loud, human constructed sounds that were beyond the range of what a human body can sustain, computer-based tools also made the process of musical invention something that could be experienced as a representation alone (or as code, another representation for those who had developed their imaginations along side these new languages). These technologies enabled the creation of artworks whose primary material was sound: sound that could go on forever, did not contain the illusion of an originary moment via inscription, could self-generate unintended material with few instructions, and for which there is arguably no original and no total author. To some extent this area of artistic production has never come into the awareness of the general public supports the notion that audio has never had to contend with a kind of "original sin" relating to Turing, we've always found ways to project human testimony into our listening to representations. Nevertheless, the arrival of computers makes it all the more important to think toward clarity in differentiating between music, sound, audio, listening, and each of their unique and overlapping qualities. This is addressed most directly in the final essay.

see page 28

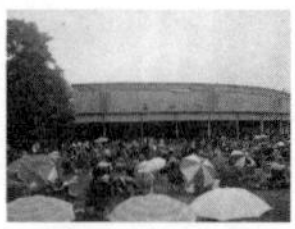

The implications for composers who were aware of these new tools were significant and schismatic. On the one hand, a composer could generate material within a range of possibilities, and conditional control structures or simple probabilistic models have remained a standard feature of contemporary programming environments that aim to produce audio. On the other hand, the authorship of this material would be in the parametric construction and if one chooses, the editing of a computer's permutations of it.

During the decades in which our relationship to audio developed and these "original-less" forms of audio emerged, media companies conscripted the public in a very careful way in order to maintain the illusion of an original and of authorship models that maintained the aura and myth of composers and musicians arriving at their music not from permutation and labor, but near-divine inspiration or genius. As the pop music of the 1980s and '90s became increasingly pseudo-diverse, radio stations consolidated, reaching a near-monopoly state in the late '90s. Computers, and by the late 1990s, the Internet, posed an existential threat to the music industry and the carefully cultivated aura of their products: a direct and ongoing elaboration of Edison's initial marketing strategy from decades prior.

Unlike the promoters who gambled in 1975 on selling tickets to proximate audio representation at Tanglewood, the record industry was not convinced that the public would pay for anything but the original mythology of albums being "music" and an item to collect as an identity construct requiring physical space—a kind of identity rather than artifacts to make space and imagination.

Girl, You Know It's True

During this time, the composer Maryanne Amacher was working on an elaborate project: a media opera for the home called *Intelligent Life*.[54] The project was never realized, but the treatments express a Utopian vision for music and a type of virtual reality in the year 2021 that contains an embedded vision for a resolution between the brain sciences, computers, artificial intelligence, and subjectivity.

In the world of *Intelligent Life*, human beings have finally grasped, as Amacher quotes Lewis Thomas, that "music is the effort we make to understand how our brains work," and that we are focused on exploring music in service of opening up our expanding our way of being. It also contains lucid insights and critique into a mouthful of a conundrum-in-hand: How can we expect to distinguish between the *what* that we listen to (e.g. music, sound, audio, language, etc.), if we can't perceive anything absolute about the air we are listening *in* with conviction enough that this perception would alter our identification of the *what*?![55]

54 All materials relating to *Intelligent Life* remain unpublished and references and documents included here are from the in-process Amacher Archive that I've been collaborating on establishing with artists Robert The and Bill Dietz.

To illuminate this with Amacher's science-fictional critique, a lengthy quotation from *Intelligent Life*, "Background to the Musical Intrigue" follows:

> **This all began with computer software for the home- designed to give people the pleasure of "composing," i.e., modifying EXISTENT patterns in Bach, Mozart, Beethoven, Reich—making subtle, or not so subtle, variations and developments of this music. What become know as 1st Order Artificial Intelligence Scores, made it possible to create "imitative" music cheaply and efficiently. By 1995 such imitative scores were toys of children!**
>
> **REALLY, any melodic patterns (rhythmic and tonal) a composer might make, and wish a variation of, could be developed in quite sophisticated ways, almost INSTANTLY by Silicon Intelligences. As fast as a TUNE could be made, it could be developed in any number of styles now, with a wonderful range of variations!**
>
> **What good was it now to be "a great composer," the ROMANCE WAS GONE. There was no longer the old joy of sitting in the studio, writing out, and recording these tunes, if 5 minutes later they could be elaborated, and developed in works of great symphonic proportions by machine intelligences! Frankly, what good was a composer's human brain, when scores for pattern re-arrangement in the many styles of music, excellent programs for "personalizing" one's own sequences, and clever novelty features, were all written. What did a composer now do? What could be composed that would not be so quickly and easily imitated and developed? THIS BECAME THE REAL CHALLENGE. Every composer soon hoped to keep "them" guessing for AT LEAST 2 weeks before his newest composition was 'DUPED'! ... This produced a GREAT EFFECT on musical thought. It resulted in the necessity for more subtle inventions on the part of composers. Or, what in fact, they were doing as human intelligences—if their next 'inspiration' (i.e., a personalized sequence of melodic, rhythmic variations) could be so easily and instantly created by "machine" intelligence?**

55 The emergence of Minimalism and Meillassoux's notion of correlationism is explored more thoroughly in my essay "Time-Based Form after Minimalism."

All of this actually ADVANCED the ART OF MUSIC, to a much higher level.

However, before this happened, composers had to face something even more DISTURBING, that writing these melodies, or "germs" to be developed—the main activity of most composers—was NO LONGER that profoundly CREATIVE, after all! It became clear that composing had not meant isolating acoustic features, beginning HERE with the SPECTRUM itself, discovering its energies and shaping. The unsettling truth was that most approaches to creating music BEGAN with EXISTENT FIGURES—melodies snatched from the great fragments of MUSICAL MEMORY! What composing usually amounted to, was a re-arranging and modifying of these patterns, i.e., OTHER MEN'S TUNES, and giving the a PERSONALIZED SEQUENCE IN TIME. And, the Silicon Composers were now doing this better and faster than they could! ... For composers to feel "worth anything" the music they now made must explore areas of sensitivity, for which NO SOFTWARE for INSTANT DEVELOPMENT yet existed! Composers had to GO IN AND LISTEN in ways NEVER DONE BEFORE! They hear, think, and explore the "unformulated," where CLUES were still insufficient for synthesis!

see page 10

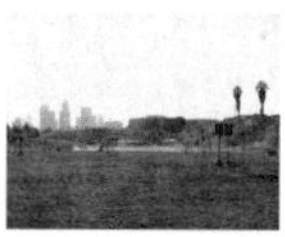

The had also to try and UNDERSTAND now, in the very deepest sense, "musical memory"—how music's many tunes, and melody traces matched "listening mind's" memory traces—rather than simply snatching these melody fragments out of the air, and re-arranging them in time. They returned to the past also, and though about some of the great early inventions in Music, which were now taken for granted. They thought about Monteverdi inventing the TREMOLO; Viadana, the FIGURED BASS. They recharged their special musical sensitivity, to INVENT.[56]

While Amacher was writing about composers, creativity, and the paradox of artificial intelligence undoing the uniqueness of what

56 These quotes are taken from the section of *Intelligent Life* titled "Background to the Musical Intrigue." The exact date of authorship is not known, but these words were likely written between 1980 and 1985.

makes human "composition" itself, these questions were being dealt with in terms of performance and authenticity by the record industry.

The curiously traumatic apex was the 1990 Grammy Awards during which Milli Vanilli won the "Best New Artist" category and was then stripped of their award when it became clear that "the artist" was not singing the songs, they were simply an image enacted to position the audio representation in a marketable way. The producer of "Girl You Know It's True," Frank Farian, to this day claims he hired them as dancers and that Fabrice Morvan and Rob Pilatus began lip-syncing during a video shoot of their own accord.[57] Among other reasons for suspicion, obvious to many music industry professionals involved with the promotion of their debut, Fab and Rob could barely speak English in 1989, never mind sing without an accent.[58] Clive Davis, the president of Arista, the label that released "Girl You Know It's True," revealed in a recent interview that Fab and Rob had a contract with Farian that gave them no royalties from the over 11 million copies sold worldwide (as of 1989).[59] Farian and Arista profited enormously. After their Grammy was stripped, Fab and Rob fell out, Rob committed suicide, and Fab released a few records, but seems to primarily be occupied with endless profiles and interviews about the scandal, most recently telling the story of their extortion on "The Moth," a storytelling program aired on National Public Radio.[60]

The technology for karaoke was refined in the same period—the late 1980s—and took hold in Asia in the early 1990s, quickly spreading to the rest of the world. While there are fundamental differences and social inversions between karaoke and secretly lip-syncing, the feeling of pretending to be amid audio is a

57 "Starproduzent Frank Farian: Deutschland Sucht Den Superstar," Spiegel Online, November 14, 2010, http://www.spiegel.de/kultur/musik/starproduzent-frank-farian-deutschland-sucht-den-superstar-und-findet-ihn-nicht-a-726090.html.

58 Marks, Craig and Rob Tannenbaum, *I Want My MTV: The Uncensored Story of the Music Video Revolution* (New York: Dutton, 2011), 362.

59 "Girl You Know It's True: Milli Vanilli," http://whitgunn.freeservers.com/Davemusic/M/milli-vanilli/girl-you-know-its-true.html.

60 "FAB MORVAN of Milli Vanilli—The Official Site," accessed May 6, 2013, http://fabmorvan.com/.

constant.[61] But even as we enacted our own imaginaries of performance in karaoke and were satisfied by the minor living room spectacle of it all, the members of Milli Vanilli were continuously shamed—we couldn't respect the scaling-up of this social architecture and the idea that pop music was a media spectacle rather than a talent show. That was only permissible for "us" and not for our media heroes.

By 2002, the recording industry had been transformed by the Internet, file sharing, and MP3, and was in a panic. One tactic to rectify the industry's loss of control within the media landscape took the form of "American Idol," an elaborate televised talent show aimed at creating music industry stars via a crowd-sourced popular vote. It entered the popular imagination (via the Fox network) in 2002. The first winner was Kelly Clarkson, whose prize was a major label record contract. Since then she has sold 25 million albums. With the show's popularity and the financial success of its winners, there now exist many similar programs, related to different labels and genres.

From where we stand now, over twenty years after the Milli Vannili incident, it is absolutely the case that a group could exist and win a Grammy award with a singer who never came on stage. In fact, since that time there has emerged an entire genre of "virtual bands," who have animated or fictional characters "stage" their shows for them. Gorrilaz, the most successful, was even awarded a Grammy in "Best Pop Collaboration with Vocals," a category initiated in 1995 that was the container for Natalie Cole's posthumous, virtualized, duet record with her father Nat King Cole. The category was eliminated in 2012.

• • •

The historical diagramming of the preceding pages locates the schizoid psychology of contemporary audio as unfolding in many locations at once. While the first visitors to the lawn at Tanglewood were contemplating whether they were in fact listening to "an orchestra" or a representation

61 We can imagine this to be the case even as we listen to recordings in a more generic sense. What is the difference between "pretending along" and "entertaining?" A stimulus constructing a shared set of social permissions that can be enacted only within its sensation present (entertainment's social consequence) versus a stimulus whereby we use perhaps the same stimulus to project ourselves as it's imaginary author.

of one, dancers in discos were euphorically experiencing monumentally louder, immersive, and more affectively scintillating auralities than ever before. Attendees to Grateful Dead shows were experiencing something related and record collectors and audiophiles were attempting to produce private experiences to recollect and produce vivid real/memory hybrids.

A knotted complex of relationships between people and machines, and a multiplicity of paradoxes in the transformation of how we trust our subjectivity confuse what "audio" has become. Its ubiquity has proved a more radical force than anticipated, and has not received the same surgical disambiguation as the image.

Audio affords an uncomfortably intimate encounter with representation, as our subjective acoustic memory, our sense of place, our personal memories, our ways of thinking, feeling, and being spark together, merging so quickly it's nearly impossible to parse the resulting composites. We can embrace all of these subjectifying forces—their difficulty and sensuality, whose value or non-value and possible relations to a pre- or non-audio "reality" are often unknown and might ultimately be unimportant.

Audio's utility is best imagined in relation to the utility of daydreams—a swimming pool, a site of poly-dimensional osmosis; learning without the necessity of verification or truth, toward justification. But in the face of audio existing within such a capitalized area of culture and our life, we will need to resist this ease temporarily to develop a relationship that captures all that can be possible in the air we can make, and its capacity to imprint schemas with revolutionary implications.

see page 15

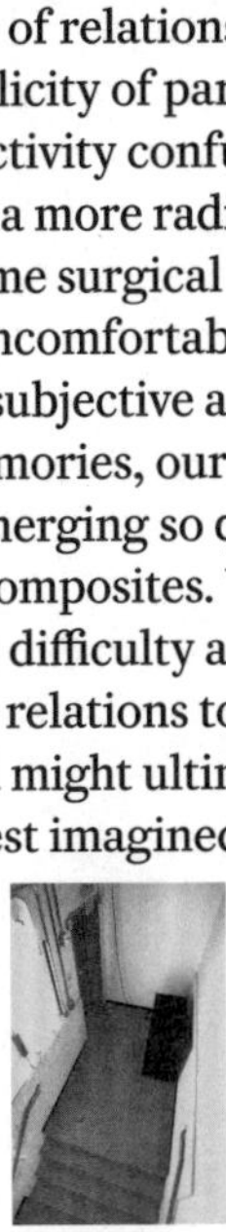

Throughout these texts we've been considering audio to be primarily a "representation," rather than a "reproduction" or "music" or "sound," and this is a critical point.

We can consider that it may be a powerful tool to understand this representation as inherently *ancestral* and as linked more broadly to the current status of subjectivity's insight into the absolute.

The potentiality of audio is in its address—of the air it produces—and how, as a place, it exists within the oscillations between our thinking, affect, our physiology, and its action. Audio are always in a productive and ephemeral reciprocity with a social architecture inhabiting it, even if we are alone. All of this is contingent upon the lexicon of memories and desires with which we ornament the presence of its transsubstance: temporaneity.

Distilled: audio as a "thing" is a perceived change in the air, the delta between a room with and without it. Within this delta are the knots I've attempted to untangle. But if we can keep this simple definition in mind, audio becomes a tool. With this tool we can make temporary places that construct a site of dialogue between sense of self, ways of thinking, ways of being alone, and ways of being together—an ephemeral social architecture with power to transform us in ways no political or philosophical solution can on its own.

The more we can learn and stimulate the role for audio that most realizes our desire for the world we want and how it feels to be there, to think there, the more we can embrace the multiplicity that characterizes our time, and the better to locate our own schematics for what will be a whole. Answers are how questions are lost to ideology.

5
Figures in Air

We will not attempt to define the primary qualities of audio, but simply continue to strip away, to move toward identifying a way of thinking that can take us to a shared language for doing so. What we are after is a model for differentiating the qualia of audio as air as social architecture that is not limited to individual subjectivity, but is within the bounds of a particular time and space, can be gathered as a temporary intersubjective absolute. We're looking for the component parts of temporaneity in order to move closer to articulating the variables modulating and intermixing in audio.

What we can achieve by doing this is to detrivialize audio as a means to heighten the importance of accessing our experience in ways that result in a reality that doesn't rely on knowledge structures beyond the intersubjective reality we hope to participate in (quantum mechanics, for example). To get there we need to filter away confusion as to the properties of the perceptual moment we are locating. We are looking for the power of audio as fiction, the constituting vessels that allow for this fiction to exist collectively much more than the power of sound as physical fact. We are here in these texts not to locate a lasting proof, but simply to practice what the Canadian artist and writer Jan Zwicky calls philosophy: thinking in love with clarity, and to follow the resonance of these forms into new, temporary spaces.

First we must eliminate the notion that audio is technology. While the quality of the air we *identify as audio* is manufactured in an ongoing way by technology, the role that technology and architecture play is *as an interface* to this air. In this view, the ideal technological and architectural interface should be more sensitive than we are, and with a means of control that affords human beings utilizing these tools to "tune" the air with as much precision and subtlety as they are capable.

These tools can improve and reciprocally "tune" us. And in this framework, every detail of the technology does impact the resultant range of possible qualias and for this potentiality, technology matters enormously for the production of audio. But even as the air is contingent on technology, we cannot mistake one for the other.

A partial list of nodes within the topography comprising the interface:

- **Ancestral sound or source duration**
- **Social and total conditions modulating the duration and its actors' acoustic self-representation**
- **Atmospherics/aurality of production location**
- **Attentional arcs projected into space and their psychodynamic feedback within the atmosphere**
- **Signal chain of capture or creation**
- **Transubstantiations**
- **Encoding or means of preparing the carrier media for transubstantiation**
- **Properties of the carrier media or virtual package (i.e. a "file," plastic, metal, glass)**
- **Possible transformations of the carrier media over time that alter the transsubstance**
- **Decoding or means of preparing the carrier media for transubstantiation**
- **Signal chain to a diffusing mechanism**
- **Spatial diffuser (a technology that places sound into space via a compressible substance: gas, liquid, or solid)**
- **The atmospherics/aurality of the diffusion location**
- **The acoustics of the location as shaped by a combination of gases, liquids, and solids subject to physics of Earth (if on Earth)**
- **Acoustic interference and intervention, atmospherics**
- **Social, historical, and total conditions constricting the**

psychophysical range of perception
- **Position of the listener's body in the space**
- **Attentional arc**
- **Spectral properties and temporal displacements of the listener's ears, torso, head, and hair**

These and related means of production are not where we will find audio, nor the capacity to model and differentiate with clarity the delta between a room with our without a particular quality of air. These qualities partially account for material contribution to a total contingency of relating to audio as air, or air as wind, with any particularity.

Attempting to locate these answers within the interface is a fearful foreclosure of thought; an impasse to finding a means of encountering what is beyond an ambivalent, totalizing relativism of imaging audio as purely original or purely artifact or purely representation. We must find a way to bypass this block by if we will hope to develop a means of identifying "figures" in the air—the schematic forms that *temporaneity* can take and the potential *puissance* of this substance's transmission.[62]

To double back briefly: audio as a thing can be located at the delta between air with it and air without it. The *something* that transfers to air across the entirety of possible interfaces and signal chains that allows audio to bloom in our imagination is temporaneity.

So to find ourselves in a time and place, however contingent, within which we can identify, savor, and construct a schema for an audio "figure," we will need to find a means of characterizing both of the above components as a resonant form, metastasizing through air and collectivizing people and imaginaries in space. A temporary social architecture made of air.[63]

62 "Figures" is a revision made in 2012 by Quenntin Meillassoux to his previous use of the phrase "primary qualities" by which he means absolute properties of things. Crucial to keep in mind is that these properties are not claimed to be "absolute" across time and space, but absolute within a framework of thinking that assumes reality to be entirely contingent (i.e. other than what we think it is or actually is in one moment of encounter).

63 Since the late 1980s the composer Maryanne Amacher discussed her work(s) as containing a collection "sound characters" whose total contingency reflects these ideas in practice. Her "characters" were not audio files or recordings, but temporary constructs in time and space, subject to the entirety of that presents facticity.

6
Sound, Audio, Physics, Imaginaries

What follows is an edited, compiled version of two talks given during the summer of 2014 at conferences held at Goldsmiths in London, and at Birmingham City University, Birmingham, UK. While some of the notions here reflect earlier texts in the book, some areas are clarified or brought into focus in the following, a year or so after the preceding texts were complete.

I am going to make an attempt to disambiguate sound from audio. I will mostly engage situated audio and audio representations of music or music-like structures because this is the lion's share of what people are talking about when they say Sound Art, or encounter it, or use the phrase.

My intention here is to sketch a partial diagram for understanding what is required to adequately situate, analyze, and work with what we refer to as Sound Art from a curatorial or theoretical perspective.

The visual arts and music both have "theory" traditions—none singular methodologies—but with some agreement as to the basic conditions of a work's existence that should be accounted for when locating *where* the work is within a piece and its context. In addition to material and airborne qualities, we need to find ways of adequately locating a work's logics of construction—the algebra of thought and action that is embedded as process and what this represents. We need to be discriminating as to whether works being called Sound Art that succeed as object of discourse are doing so because they rely on models that would be uninteresting if not relocated to this new frame, in particular.

We don't yet have this in the 2014 Sound Art wave, we didn't in 2000 Sound Art wave, and we can assume from the results that we didn't have it in the 1960s, either—though at the time one could just say it was work by a composer, as Alvin Lucier has been able to maintain: he's of the last generation of which that claim remains possible. What happened? Where did all the "composers" go? For lack of a better idea, let's blame Warhol's emphasis on the social image of things and his instrumentalization of The Velvet Underground, or perhaps we can blame Donald Judd's oedipal rage? There are several interesting historical lines we could trace, but instead I think we can look to ourselves and what frameworks we have and which we lack, none are mysterious to learn. But this learning will not only come through texts and discussions—there is the inescapable labor of listening through history and training ourselves in this regard as well. Where and how this happens might determine a lot for listening in the arts.

And while there will always be many ways to analyze and come to some curatorial perspective on a work—and I am in no way hoping to foreclose on idiosyncrasy—we should learn from the many defunct art movements or sub-disciplines that we no longer discuss except as period pieces or esoterica. Historically Sound Art (as with most forms of aparadigmatic experimentalism, not only in art) operates as Marx described his mole, popping up, disappearing, the questions or challenges the area of practices pose sleep and then awaken to engage the culture another day with hopes of a new resonance. And like any transition among power structures and tastes, continuity and traction are elusive but ultimately critical to sustainability, thus the obsession with integrating into the areas of culture where the most social capital/gravity exists.

It's worth saying that this feeling of partial exile is widespread and not limited to truly esoteric practices like Sound Art: even Jay-Z feels excluded from the broader discourse of the arts and has made recent attempts to argue for his area of practice being considered equal to that of Great Works of Art (his *Picasso Baby* composition and video featuring art world celebrities is a concise example). The frame of pop domination turned for him into a demeaning badge of low-culture (which perhaps misses the current multiplicity between high/lo-/no-/other-brownesses. So, there are many areas of culture in some state of diaspora in relation to the most-difficult-to-obtain forms of cultural capital.

While it isn't my concern to think through the power structures in play here, the traction I am looking for it simply that continuity, influence of ideas, sophistication and nuance of discourse are interrelated, and if a more advanced take on listening, audio, and sound is to take hold in the culture, Sound Art is actually a reasonable launch site. So the goal here is to propose some basic beginnings to a framework, that while seemingly simple, I assure you are needed in the field. Assuming it can dissolve its knowledge into the broader fields, Sound Art has the potential to commit a meaningful disciplinary suicide, by proposing terms and a loose theoretical basis that can allow a more powerful and legible relation to sound and audio in the arts, generally.

So the question here is one of criteria and of clarity in thinking—starting with the most basic terms in play and trying to make clear distinctions where possible. This is important as the basis for a continuity, for sharing and extending the possible meanings of works made within the field, and without negating the neurodiversity of a notion like quality. We need to establish some basic footing for having the option of discussing quality, with more and better questions on the table.

Sound

In one sentence: Sound is the domain of physics.

It is a generic category for compression waves through a medium. And even though, to some extent, our physiology has a relationship to sound as a sensation, it has nothing to do with what we can or cannot hear, or sounds we can or cannot make. Sound is a scientific term, and we have technology to roughly measure sound both within and outside of the provenance of our senses. Importantly, these tools are not seeking meaning, but attempt to quantify and via quantification make sound an intellectually operable "thing," which has largely been accomplished to support various dimensions of engineering. As with music notation, it's clear the abstractions we are capable of making do not have the resolution nor account for the necessary qualities needed to believe we're representing the most meaningful aspects of sound for our perception.

For thinking about sound in art it is imperative that we keep this in mind: that while accurate measurements of sound remain "good enough" for operating on it toward many practical applications, they in no way *yet* reflect the true, known, complexity of the phenomena. For example, computational models used in acoustic design still use ray tracing to simulate sound's movement in space and in combination with materials in the form of reflections, absorptions, diffraction, etc. When, in fact, as we know, sound is not a ray, but a complex, often chaotic, wave. These models remain beyond our current personal computing power—and according to Ben Markham, an acoustician for Accentech in Boston, attempts remain within the domain of supercomputers and speculation.

So when we draw sound or use software related to multichannel composition of some kind, we draw paths—and these paths do not represent sound, but merely a reductive, easier to comprehend subjective localization of it. Yet if we could see sound bouncing around a room, we'd understand that it is a complex four-dimensional form whose characteristics for self-interference, rupture, and unpredictability are magnificent. Until we choose to engage these more detailed aspects of the phenomenon and develop operable subjective impressions, sound is the domain of physics.

It isn't easy to be a sound artist by this definition. As a thought experiment let's imagine for a second how music history and theory would be different if all along we could have *seen* the acoustic energy moving through space as complex multidimensional forms? Very quickly composers and the public would have become involved with this dimension and there would be a long history already of combining this *graphic listening* with other forms, be it what is demanded of us listening to romantic narrative constructions of orchestras or techno at Berghain. Imagine how we might approach landscape design or office cubicles. Instead, we built light shows in discos, control the staging and costuming of music in order to make what is invisible, visible, or shape bilaterally, both—in an expanded sense we could consider these acts as psychoacoustic! Mostly the field of acoustics, in a practical sense, can be reduced to curating absorption.

We should feel free to use audio, music; any acoustic means as a tool for knowledge of areas we've foreclosed to science. At the same time we should be aware of how simplistic demonstrations of quite mundane scientific fact is perhaps performing free labor for the false sense of science's grip on knowledge of what this world and it's things are. The tacit use of quantitative implications within a work is insidious if we think of them in this light.

There have been a few remarkable "sound" artists to meet the complexity of the physics with unconventional research methodologies that honor the potentiality of what can't yet be measured. People who use sound as a tool to explore speculative edges of physics and cognition—this is where we would find a true Sound Artist by this definition and where I locate the work of Maryanne Amacher, for example. Her work is most profoundly understood as being in large part an approach to highly speculative acoustics dealing with chaotic spatiality formed from extremely subtle use of reflection. She would joke that it was "crackpot acoustics" but she was entirely serious and those who experienced her work and witnessed her process of slowly establishing a sound system within architecture, matching spectral materials to clusters of loudspeakers—her work was experimental in a very strict sense of the word—and to work with the complexity of contingent variables present in any sound work requires this ethic.

Yet to say that sound, in this definition, is a "material" for art is again a grotesquely de-subjectifying position for an artwork. I like to read these questions as a strict Piagetian—I believe in the more esoteric notion of his term "schema," which asserts that we accumulate schematic impressions/learnings of how "things" work—their relations and interconnections—and that the content of these impressions is secondary to the dynamics of the structure observed. Art can be thought of productively as about learning in a Piagetian sense, and in this frame the material of art is not sound, not paint, not objects, but the schematic play we perform (with body, attention, our entirety) in order to absorb the work. For time-based art this seems to have a lot to do with entrainment, but not uniquely so. For this reason to discuss Sound Art as if it were frozen representation, purely semiotic, purely phenomenological, purely physical, etc. is a false binary and an intellectual failure to perform the more open, multivariate, subjective engagement that makes art not only design or a correlation chain.

Sound can be a tool with which to shape material—as a baton controlling a marionette, but the puppet is our attention to sound as a listener even if this means that our bodies are changed in a chain reaction beyond light speed; not the speed of sound, but the speed of consciousness, whatever that might be for you. There are no sound objects, only listening objects, and while sound is, in fact, a wave, for our purposes as speculative non-scientists, it is but the carrier wave for modulations coming from a complex of other possible materials that build an impression, a schema, that can receive the full riches of our subjectivity and model *alternatives* via listening.

Audio

I will take more time with this category as the vast majority of Sound Art can be read as functioning primarily as the staging or situating of audio and increasingly audio is a component of artworks not being read within the history or context of audio.

I am trying to articulate the answer to the question of what travels through audio processes, what is the *enough* or the *something* that makes it a convincing illusion. I am making a case for this something being a compound mined from place that I will call *temporaneity*, and I don't mean to pretentiously make up words, but as best I can tell we don't have words to hold the concept of the kind of cyborgian codicil for time that we must if we are to understand current and recent technology relying on the quantifying of time or in this case encode, hold, and represent as audio.

• • •

And while we go into this spiral of what is referred to in other situations as a "signal chain," we should keep in mind that while audio processes can have many phases of transubstantiation—in the final phase of the process, one enacts an *instance* of audio, and audio processes require this of us or someone, always—it results in a temporary place, a temporary social architecture whose substance is not sound or the transubstance we are sleuthing for, but *air* and this air is only partially constituted by sound and only partially encountered with the ear.

Some assumptions I should disclose first:

1. **That the inclusion of audio in an artwork is a suggestion from the artist that we should listen to it and that this listening requires of us a durational engagement.**
2. **That the audio's content, when listened to, can be read not only for its correlational value within the conceptual network of the piece, but unless barred or encouraged in some way, within the history and aesthetics of a relevant area of audio practices.**
3. **All audio exists as a temporal form and like any form, it can and should be read and isn't a transparency or cultural neutral.**
4. **Every artwork is situated among sound, whether it is put there by the artist or is just in the air.**
5. **Audio is not music but in certain situations audio can become music, but mostly audio represents musical structures that can produce affective results in an overlapping but distinct register as music.**
6. **Radio and telephony is not audio, but a distinct psychic typology or lexical range of listening activities.**
7. **Modes of representation are not interchangeable. One speaker with another, one room for another, headphones for another pair of headphones or a PA.**
8. **I am consciously not addressing edge-cases here and so there are many caveats and snags here in the name of brevity.**

In one sentence: Audio is not a technology, but a process that employs technology to construct temporary social architectures made of air. These representations' primary quality is a temporaneity of no origin, which nevertheless we treat as human testimony. *If audio is a thing, we can find it as the delta between air with and without it.*

This process can have many phases of what may be most accurately called transubstantiation—but in the end, when one enacts an *instance* of audio, and audio processes requires this of us, it results in a temporary place, a temporary social architecture whose substance is not sound but *air* and this air is only partially constituted by sound and only partially encountered with the ear.

It is uncanny that audio perceptually functions at all. In a slightly alternate universe listening to audio could have been similar to a cat looking into a mirror. The cat will not believe the reflection to be real or related to it, the cat will not recognize the representation of itself in the mirror as a representation even. When we record our own voice and play it back we may feel tremendous alienation from the sound, but also know that it is not someone else. But what is on the tape or in the MP3? What is in the air? What is the "enough" that audio achieves so readily as to set our imagination in motion to fill in the details?

The sub-technology that is the keystone of all audio technologies and can't be removed is an underlying periodicity or clock. Whether it is an electrical frequency that is modulated, a spinning wax cylinder, a record lathe, or a D/A converter, there is always a method to quantify and to maintain this quanta's logic throughout the system. The one component of any audio system that, if degraded, will trash the capacity for our imagination to fill in the details is this property of clocking. In digital audio this is the sampling rate—you can crush the bit depth to almost nothing and still find some resonance with the original character of audio with all of its bits filled. With a CD there is a laser reading at a pulse of 1/75th of a second as an additional overlay. Crush the sample rate or add chaotic periodicity as in Yusanoe Tone's work and the illusion no longer applies and we become the cat in the mirror. Tone's work within this frame is not audio, but music, performance, where he breaks the mirror and a machine becomes something else.

Audio really should not pass this Turing-like test. That we can imagine so much and think and feel through audio seems improbable and implies we are entirely and sweetly gullible consumers of memory materialized. Yet it does and has from the beginning of the technology where the media involved were capturing almost nothing except a trace in time. But this clear temporal trace is the enough for us to believe it is testimony to a humanity that it does not contain *unless* we do truly believe in transubstantiation.

A temporary formula for *Temporaneity*, to be very concrete about what I mean:

Temporaneity is the reciprocal of your sense of place with the change in a listening arc /entrainment episode subtracted plus your associational inventions as multiplied by the sensuality of timbre to the power of your memory shadow recalled.

This or something along these lines is the trans-substance of audio. There is sound in the air, the earth, the water, or from nothing in software—the affective and physical qualities of these things *in place and as constituents of place are not possible to displace*. There is no capturing of this *there* in its entirety. Audio, photography, videography, cinema: these are like fracking devices and violently mine with great increasing sophistication various *somethings*. Our attention moves with our senses, we don't just mindlessly digest aurality, we build it with ongoing analysis—we compare arrival times of sound at orders of magnitude unthinkably more precise than we see color or smell chemicals. Durations are bracketed by our attention. When several years back cochlear implants became possible it was discovered just how vast our aural memory is—for timbre, for differentiating acoustics, for localization—the sensation of timbre is like touching fabrics of infinite possibility association and timbre are inextricably linked—if yesterday you had a cochlear implant you would, literally struggle to differentiate a cat from a helicopter. This recognition leads you into the shadows of your memory, recalling in extreme rapidity overlays of previous sensations, decision, feelings—*all of these factors are part of what can be contained within audio's transubstance* and it seems like only a human being or a humanness could produce such a dynamic response. Yet there is nobody on the other side of the phone: it's just us, daydreaming all of this to life, even to the sound of a scratchy wax cylinder with almost nothing there. This fact, that nobody is on the other end of the line, we need to take stock of this, because when we hit play, we're imagining someone there: John Coltrane is dead. You never even experienced his presence. But audio, while it may be enough, it is not an old *is*, it is a new one that you make for yourself, like counting hits on a website or imagining the visitors to your museum show—we are having a conversation with ourselves.

So perhaps some examples are useful, some also an exception to the rule: We are all here, holding each an object whose function is a temporal displacement. Let's say it is a reel of magnetic tape or a shellac disc we just lathed—as tangibly physical as audio has ever been, and no longer is. The metals of the tape have been shaped just moments before along the path of our own voice, you just recorded yourself singing in the shower or performing *I am Sitting in a Room*, the disc cut into a topographically ornate spiral of valleys that trace the movement on our breath made by our vocal apparatus and the room. Are we now holding our *voice* in this time and place in our hands? Is it just tape or just shellac? If we aren't holding our voice in our hands, yet we can enact a representation of it momentarily, what do we hold? We have to dispose of the language of the illusion. We could say we have a "recording" or we have an "audio file"—this is, more or less the language of consumption at this point. We need an alternative.

• • •

So this tape we hold in our hands contains a *something* of our voice. We can likely agree that the synesthetic impression or sense of *duration* we experienced while making the recording has entirely disappeared. This *duration* is not your stopwatch but what we felt, how long we felt it was, the informational crosstalk of our other senses that influenced our time-sense during the recording (the temporality), the other presences in the room, the moments that preceded its effect, how permeable or impermeable a space, where are the walls in this room with no walls that is time?

This aggregate temporality sensed in the moments passing and remembered as an episode is what I mean by *duration*. Somewhere in Bergson's neighborhood. This experiential unit has nothing but a correlational, linguistic relationship with the quantization of durational episodes by a clock of any audio process, not only philosophically, but also practically. When I say duration I mean an actual time-space impression, richly perceived but beyond quantification. Contingent, sure, but its contingency doesn't negate our attachments or specificity of value and storage location.

But in thinking about audio, duration for us need not be limited to the clearest case of recording places and their contents in time; it is also the shaping of sound into a duration by a human being that accomplishes the same thing for us as we enact it in air at another time, another place. So in this view, the audio production studio, in any form, is a means through which synthetic durations are composed. A sense of *there* is constructed and shaped over many episodes of *there*, aggregated into a composite temporality, objectified so it can then be represented in the future via playback. Audio has been and always will be a fantasy beyond death.

• • •

So now, back to the voice-like thing we might or might not have in our hand—as we listen now to our tape and observe the relationship between our memory of the original duration recorded and our un-clocked sense of it via the audio representation, we are left with a somehow related, but additional impression of the duration or an originary moment the recording is testimony to. In our most lucid moments we might experience this is an overlay and grasp the difference with some precision, or with listening training learn to do this, but quite likely is cause for what Jean Piaget called the *deforming assimilation*. To protect our the stability of our ego we deform the world to lexically slide into the sameness of our experiences rather than to understand deeply the difference and construct new schemas for what is possible based on ongoing subtle difference; the learning we can do in every instance.

So we are mutating not only the present but also our memory in this act of listening. And as we listen, likely we remember not only the duration of the inscription as an overlay, but others overlay as well. Here we can again find Bergson and consider how little choice and control we have in which memories we recall and for what purpose our mind identifies them as salient information for informing the present. This difficulty in We aren't wonderful at controlling what we remember when we listen, where our senses go—most of us can barely meditate for a minute holding a single word or thought without interruption of some memory, anxiety, or curiosity.

During playback this sensual crosstalk is a new constellation, we can't remove it from our experience. Even in an anechoic chamber, a sensory deprivation tank—there is no escaping the aggregate sensation of being where we are to some extent. But nevertheless, if the tape machine was calibrated accurately, we know that a quantification of the originary duration is being represented during playback with as much accuracy as was possible during the inscription. That quantity is placed into a new time, making a new episode of durational experience, and has regardless of context a spell-like quality that can be distinguished as a constant amongst other perceptions.

• • •

Let's take another modulation: as we listen online to a "real-time" feed of a microphone placed deep underwater—let's imagine something great like the oceanic waveguide where sounds travel around the world, or in a department store, a live concert stream, or a webcam of someone's garage full of illegal snakes: in these cases are we to believe that nothing *originary* is transferred across time and space? If that were true we would have trouble fantasizing about how many snakes. What do we believe to know about these places by listening via the remote microscopy? And maybe more important, why do we desire to believe these things against so many factors that would dissuade us if we compared these beliefs in-depth to our knowledge of the mediations present?

In the first case, we could not be there listening in the same way—with scuba gear we would hear our breathing or in a submarine, we would hear the machine and listen through another microphone, but be there in time. For the others, what is the difference even in how we imagine listening as we read or listen to a description, in this moment, either by being there or by being elsewhere, imagining this other "there." Our experience right now, projecting these examples, can tell us a lot about these differences- and in many ways, text when used as in this paragraph can be quite close to the procedures of audio within our imagination.

Yet the obstacle to developing a resolution to this possible similitude of originary testimony to a human presence and representation is that we can never be in both locations at once to compare. The differences and the slope of these discontinuities in meaning with the control of our own subjectivity, however contingent, are not possible. No human has experienced this, even via, say, astral travel or the reports of CIA remote viewing experiments. But audio's intersubjective basis, our implicit social contract around how it means, is clear: we believe that there does exist some sense of testimony in the experience of audio that transfers not the entirety of the originating situation's qualia or even acoustic reality, but something of it.

So now back to the notion of an analytic framework: if as I am proposing, audio's basic feature is this highly charged fossilized and represented temporal arc of no origin in a reality we can know, then one required question when reading audio from an analytic or curatorial perspective is: How does the audio use this clock?

And then the next would be: What does this temporal logic, an overlay upon the particular signal chain's underlying pulse, represent and where are these structures or logics drawn from? How is their mode of representation functioning within the overall schematics of the work? How is the total contingency of audio leveraged or not for the overall impression of the work's place? How has it changed the air?

This seems entirely basic, but is rarely a component of the discourse around artworks that are objects for listening in time. Since minimalism one could argue it's been a difficult conversation to have outside rarified, academic circles continuing the modernist experiment in music. So for better or worse, the apex of this discourse on schematic dimensions of time-based art was perhaps during the Great Form Wars of the early to mid-twentieth century in classical music that bled into cinema, dance, and performance more broadly. This has all gone out of vogue until very recently, when I see indications that the ban on criticality around form in the time-based arts—and a total acquiescence to mass media in the form of what has recently been called Poptimism—is fading just a bit. We can hope that the next wave of Sound Art (hopefully simply called art this time) will take part in the reintegration of a more detailed exchange around these questions.

Of course the clock is not the only required frame. Audio representations always have an inescapable referentiality or language-like property via timbre in an expanded sense of the word. And from a misunderstanding of the semiotic and schematic meanings contained in the collision of the temporal and

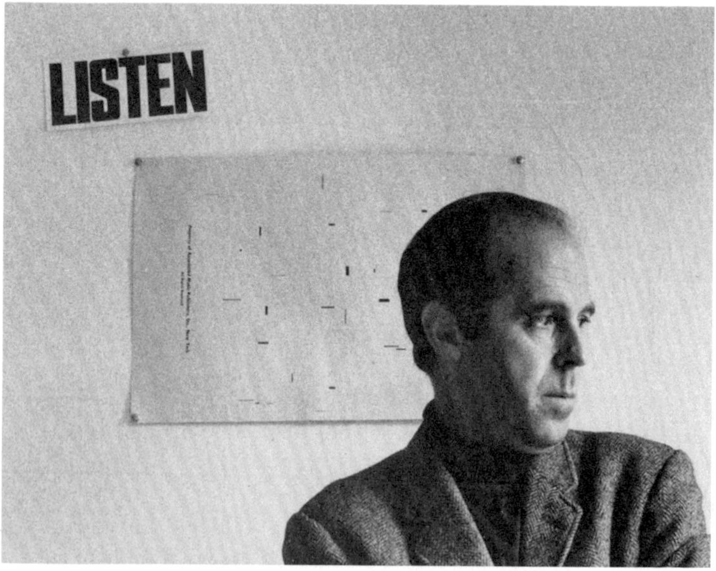

Earle Brown posed in front of an early score from which his slow intervention into the collective habits of classical musicians' togetherness would unfold.

the significant, various straw men have popped up sound studies the send us backwards to oedipal debates between Donald Judd and Clement Greenberg. We can't escape that audio is emotional, historical, racialized, sexualized, gendered, geographic. Audio is an ideal container for the multiple battles over essentialisms are long gone, we can put them to rest and embrace the broadest most delicate scope we can and audio's potential for embracing them all in great detail.

There are other lower level frameworks to account for as well, each with its own literature, analytic frame, and specialized language. If we are to fuse Sound Art with its practical and aesthetic siblings, we need to develop a way of differentiating between attentional strategies from documentary, diegetic sound design in film, storytelling, audio synthesis, cultural history, social work, and many other forms.

So when we encounter a work, what language do we use to describe the audio? Are we looking for work that has a logic of construction familiar to our existing expertise? Are we looking for bulletproof post-conceptual art correlationary crystals? Do we have the necessary tools to analyze the audio in its context and to draw a connection between it and the other elements of the work, often sculptural, textual, a reading of a site, an obscure archive? Are we easily tricked by modulating the context of one temporal or timbral strategy to an audience lacking the tools to approach the work with the same criticality we would expect from our own particular area of expertise?

A large percentage of Sound Art can be located as creative staging or situating of audio. And more than any other category, the audio often contains representations of music or music-related temporal structures. Some are appropriated and used purely as reference, but often Sound Artists are escaped composers or formerly musicians. Susan Phillipsz's work is the most recent and widely acclaimed example where it is difficult to claim a space for it as beyond music in terms of why her pieces work, why they are enjoyed, what attracts people to them and what they deliver. And if this is the case, we need to think about what the music represents ethically, how it is situated in the culture, and allow that into our perception of the work's mode of significance.

From here a rapid fire series of questions and observations:

As a field of thinkers about this area of artistic practice, why are we so eager to distance ourselves from music and the modes of engagement and responsibility that less distance would imply? In this conference we hesitate before saying "music" as if it might disqualify us or be a misdemeanor. Why have we shied away from fighting for the vast definition of music that was in play during the 1960s and achieved within the mainstream of classical music, for example? We aren't claiming this history and sharing it—we tend to bury these sources for our practices. Yet Hans Ulrich Obrist just published a volume called *New Music*—and so why is Obrist claiming this history now? Why would he claim it at the same time Sound Art isn't? Most likely it is to find a visual artist openly admitting the influence of Xenakis or Stockhausen in a wall text than a Sound Artist. Museums tend to present this work as if it spontaneously generated, and quite often several pieces within an exhibition of sound art are nearly verbatim replicas of previous works, unknown to the curator.

If a work contains audio, that does not make it Sound Art on its own, correct? Why would we call a CD release of a compilation of works of Sound Art? Is it because we have reflected on the temporal/timbral information and determined that it is simply NOT a representation of musical information we might understand from somewhere else? Are we just an alienated music diaspora? Does anyone here want to fight for the next thirty years to force a distinction between Sound Art and plain old Art? Do we see sculptors fighting to be called sculptors or are they just artists? Do we call people "video artists" anymore? Do we believe that there should be the R&D center for listening somewhere? As has happened at a recent art fair, are we comfortable with the idea that audio played in a limousine for VIP visitors makes the work Sound Art and not music being played in an automobile like every other car? Is this where we want to find our work? If you were the VIP drinking champagne in the limo, would you be curious to discover a new way of listening? Can you barely wait to make a piece for a stairwell, elevator, or headphones? What is at stake here beyond opportunism, rebranding, relabeling?

There has always been a listening component to art experience. Music is art, is it not? Music and performance are compositionally visual, are they not? So where is our traction? Why not join the post-medium discourse and feel free to simply be artists who happen to think through music or audio or performance or object making?

• • •

Audio's particular opportunity affords an uncomfortably intimate encounter with representation, as our subjective acoustic memory, our sense of place, our personal memories, our ways of thinking, feeling, and being, the social rules provoked by the audio—all of this sparks together forming so quickly it's nearly impossible to parse the resulting composites. Before the Walkman, audio was almost entirely social—and this "air architecture" contained an even more chaotic set of forces and possibilities. Listening requires a lot of time and attention—it's difficult—and we can't always do it. Listening and finding a relationship to a peculiar air is contingent. We can embrace all of these forces as audio's subjective grace.

Audio's utility is best imagined as in relation to the utility of daydreams, a swimming pool, a site of poly-dimensional osmosis; learning without the necessity of verification or truth toward justification. But in the face of audio existing within such a capitalized area of culture and our life, we will need to resist this ease temporarily to develop a relationship that captures all that can be possible in the air we can make, and its capacity to imprint schemas, ways of thinking, systems of order, possibilities. At a minimum art should support Gumbrecht's very elegant definition of consciousness: "an awareness of alternatives," and perhaps with this ethic of utility and a broader set of shared tools, curators can help to heal some of the historic fractures and advocate not for Sound Art as if it were an ethnic-themed group show, but a more perforated relationship to both art and to music.

The potential of audio is in its address—of the *air* it produces—and how, *as a place*, it exists within the oscillations between our thinking, affect, our physiology, and its action. Audio are always in a productive and ephemeral reciprocity with the temporary social architecture inhabiting it, even if we are alone and all that surrounds us is air.

Figures in Air
is published by
Inventory Press, LLC
2305 Hyperion Ave.
Los Angeles, CA 90027
inventorypress.com

Editor:
Eugenia Bell

Design:
IN-FO.CO

Printed and bound in Singapore
by Pristone
Second printing
ISBN: 978-1-941753-01-9

Distributed by:
ARTBOOK I D.A.P.
75 Broad Street, Suite 630
New York, NY 10004
artbook.com

Front cover:
Concept and realization: Micah Silver
Vapor modeling: Nate Hess
http://natehess.com
Photography: Tim Wilson
http://timwilsonphotoworks.com
Model: Emilie Baltz

Back cover:
Toby Old
tobyold.com

Image credits:
p. 1: Courtesy of Douglas Self; **p. 2–7**: From the collection of the Margaret Herrick Library, Academy of Arts and Sciences; **p. 8**: Courtesy of Experience Music Project; **p. 10**: Courtesy of Daniel Teige; **p. 12**: Courtesy of D.A. Pennebaker; **p. 14**: © Archive of the Stockhausen-Stiftung für Musik, Kürten (www.karlheinzstockhausen.org); **p. 15**: Courtesy of Sergei Tcherepnin; **p. 16**: © Richard Pechner/rpechner.com; **p. 18, 82**: Courtesy of Toby Old, www.tobyold.com; **p. 20**: Copyright Institut National de l'Audiovisuel; **p. 22**: Courtesy of Dave Rat; **p. 24**: Woodstock 99. Photo credit: Jay Greinsky. Liscensed via Creative Commons Attribution; **p. 26**: Orson Sieverding, HORST KRZBRG, Berlin, 2010, © Orson Sieverding, VG Bild-Kunst; **p. 28**: Yo-Yo Ma at Tanglewood on August 9, 2009. Photo credit: Doug Orleans. Liscensed via Creative Commons Attribution; **p. 44**: Courtesy of Erich Malter; **p. 58**: Courtesy of Micah Silver; **p. 67, 68**: Courtesy of Chris Jaffe; **p. 73**: Courtesy of DJ Punch; **p. 74, 76, 77, 81**: Courtesy of Richard Long and Associates. Reprinted with permission from: Fierstein, Alan and Long, Richard. "State of the Art Discotheque Sound Systems: System Design and Acoustical Measurement." Lecture, Audio Engineering Society's 67th Convention, New York, NY, Oct. 31/Nov. 3, 1980; **p. 78**: Courtesy of Michael de Benedictus; **p. 108**: Courtesy the Earle Brown Music Foundation